McGraw-Hill's

500 Statistics Questions

Also in McGraw-Hill's 500 Questions Series

McGraw-Hill's 500 American Government Questions: Ace Your College Exams
McGraw-Hill's 500 Calculus Questions: Ace Your College Exams
McGraw-Hill's 500 College Algebra and Trigonometry Questions: Ace Your College Exams
McGraw-Hill's 500 College Biology Questions: Ace Your College Exams
McGraw-Hill's 500 College Chemistry Questions: Ace Your College Exams
McGraw-Hill's 500 College Physics Questions: Ace Your College Exams
McGraw-Hill's 500 Differential Equations Questions: Ace Your College Exams
McGraw-Hill's 500 European History Questions: Ace Your College Exams
McGraw-Hill's 500 French Questions: Ace Your College Exams
McGraw-Hill's 500 Linear Algebra Questions: Ace Your College Exams
McGraw-Hill's 500 Macroeconomics Questions: Ace Your College Exams
McGraw-Hill's 500 Microeconomics Questions: Ace Your College Exams
McGraw-Hill's 500 Organic Chemistry Questions: Ace Your College Exams
McGraw-Hill's 500 Philosophy Questions: Ace Your College Exams
McGraw-Hill's 500 Physical Chemistry Questions: Ace Your College Exams
McGraw-Hill's 500 Precalculus Questions: Ace Your College Exams
McGraw-Hill's 500 Psychology Questions: Ace Your College Exams
McGraw-Hill's 500 Spanish Questions: Ace Your College Exams
McGraw-Hill's 500 U.S. History Questions, Volume 1: Ace Your College Exams
McGraw-Hill's 500 U.S. History Questions, Volume 2: Ace Your College Exams
McGraw-Hill's 500 World History Questions, Volume 1: Ace Your College Exams
McGraw-Hill's 500 World History Questions, Volume 2: Ace Your College Exams
McGraw-Hill's 500 MCAT Biology Questions to Know by Test Day
McGraw-Hill's 500 MCAT General Chemistry Questions to Know by Test Day
McGraw-Hill's 500 MCAT Physics Questions to Know by Test Day

McGraw-Hill's

500 Statistics Questions

Ace Your College Exams

Sandra Luna McCune, PhD

New York Chicago San Francisco Athens London Madrid
Mexico City Milan New Delhi Singapore Sydney Toronto

1 2 3 4 5 6 7 8 9 10 11 12 13 14 15 16 17 18 QFR/QFR 1 0 9 8 7 6 5 4 3

ISBN 978-0-07-179872-3
MHID 0-07-179872-2

e-ISBN 978-0-07-179873-0
e-MHID 0-07-179873-0

Library of Congress Control Number 2013936206

Illustrations by Cenveo

This book is printed on acid-free paper.

CONTENTS

INTRODUCTION

You've taken a big step toward success in statistics by purchasing *McGraw-Hill's 500 Statistics Questions.* We are here to help you take the next step and score high on your first-year exams!

This book gives you 500 exam-style multiple-choice questions that cover all the most essential course material. Each question is clearly explained in the answer key. The questions will give you valuable independent practice to supplement your regular textbook and the ground you have already covered in your class.

This book and the others in the series were written by experienced teachers who know the subject inside and out and can identify crucial information as well as the kinds of questions that are most likely to appear on exams.

You might be the kind of student who needs to study extra before the exam for a final review. Or you might be the kind of student who puts off preparing until the last minute before the test. No matter what your preparation style, you will benefit from reviewing these 500 questions, which closely parallel the content and degree of difficulty of the questions on actual exams. These questions and the explanations in the answer key are the ideal last-minute study tool.

If you practice with all the questions and answers in this book, we are certain you will build the skills and confidence needed to excel on your exams.

—Editors of McGraw-Hill Education

CHAPTER 1

Classifying Data

1. Which of the following illustrates quantitative data?

(A) foremost colors (red, yellow, orange, or other colors) of flowers in a garden
(B) sex (male or female) of users of a website
(C) lengths (in meters) of broad jumps
(D) satisfaction ratings (on a scale from "not satisfied" to "very satisfied") by users of a website

2. Which of the following illustrates quantitative data?

(A) ten-digit US social security numbers of students in a classroom
(B) college classifications (freshman, sophomore, junior, senior)
(C) flavors of ice cream
(D) weights (in pounds) of newborns at a regional hospital

3. Which of the following does NOT illustrate quantitative data?

(A) freezing temperatures (on a Celsius scale) of chemical mixtures
(B) heights (in centimeters) of plants
(C) survey responses (on a scale from "strongly disagree" to "strongly agree") of likely voters
(D) number (0, 1, 2, or a greater integer) of people attending a conference

4. Which of the following illustrates qualitative data?

(A) eye colors (blue, brown, green, hazel) of babies
(B) distances (in miles) traveled by students commuting to school
(C) heights (in inches) of students in a classroom
(D) number (0, 1, 2, or a greater integer) of students absent from school

5. Which of the following illustrates qualitative data?

(A) number (0, 1, 2, or a greater integer) in favor of school uniforms
(B) names (first and last names) of students who took an exam
(C) ages (in months) of a class of preschool children
(D) costs (in dollars and cents) of tuition at four-year universities in a particular state

6. Which of the following does NOT illustrate qualitative data?

(A) ten-digit US social security numbers of students in a classroom
(B) numbers on football jerseys
(C) heights (in centimeters) of plants
(D) customer ratings (on a scale from "very unpleasant" to "very pleasant") of an online-purchase experience

For questions 7–10, determine the highest level of measurement that can be attributed to the data set described.

7. Eye colors (blue, brown, green, hazel) of babies

(A) nominal
(B) ordinal
(C) interval
(D) ratio

8. Distances (in miles) traveled by students commuting to school

(A) nominal
(B) ordinal
(C) interval
(D) ratio

9. Number (0, 1, 2, or a greater integer) of students in a classroom

(A) nominal
(B) ordinal
(C) interval
(D) ratio

10. Survey responses (on a scale from "strongly disagree" to "strongly agree") of likely voters

(A) nominal
(B) ordinal
(C) interval
(D) ratio

11. Which of the following levels of measurement generally is/are considered qualitative data?

I. nominal
II. ordinal
III. interval
IV. ratio

(A) I only
(B) I and II only
(C) III only
(D) III and IV only

12. Which of the following levels of measurement generally is/are considered quantitative data?

I. nominal
II. ordinal
III. interval
IV. ratio

(A) I only
(B) I and II only
(C) III only
(D) III and IV only

13. Which of the following illustrates discrete data?

(A) number of brown-eyed children in a classroom
(B) distances (in miles) traveled by students commuting to school
(C) heights (in inches) of students in a classroom
(D) survey responses (on a scale from "strongly disagree" to "strongly agree") of likely voters

14. Which of the following illustrates discrete data?

(A) number (0, 1, 2, or a greater integer) in favor of school uniforms
(B) initial weights (in pounds) of participants in a weight-loss program
(C) ages (in months) of a class of preschool children
(D) weights (in ounces) of boxes of cereal

15. Which of the following illustrates continuous data?

(A) number of students in a classroom
(B) number of patients who improved after receiving an experimental drug
(C) heights (in centimeters) of plants
(D) number of customers who rated an online-purchasing experience as "very pleasant"

16. Which of the following illustrates continuous data?

(A) ten-digit US social security numbers of students in a classroom
(B) number of female users of a website
(C) lengths (in meters) of broad jumps
(D) satisfaction ratings (on a scale from "not satisfied" to "very satisfied") by users of a website

CHAPTER 2

Organizing Data

17. The table shown displays the results of a poll that asked 240 college students to identify their favorite movie genre. If a circle graph is constructed using the data in the table, what central angle should be used to represent the category comedy/romantic comedy?

Movie Genre Preference

Genre	Number of Students
Action/adventure	33
Drama	16
Comedy/romantic comedy	76
Musical	11
Mystery/crime	12
Horror/suspense/occult	48
Science fiction/fantasy	35
Other	9
Total	240

(A) 32°
(B) 76°
(C) 114°
(D) 246°

18. The circle graph shown displays a budget for a monthly income of $3,500 (after taxes). According to the graph, how much more money is budgeted for rent than for food and clothing combined?

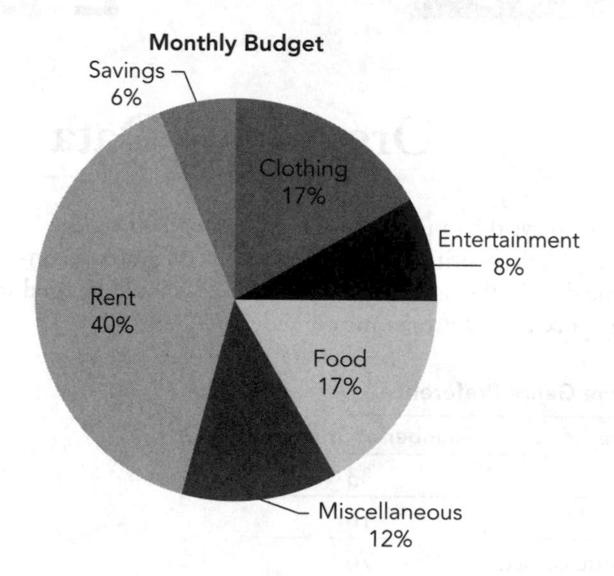

(A) $210
(B) $595
(C) $1,190
(D) $1,400

19. According to the pictograph shown, how many of the 40 cat owners surveyed responded "yes" to the question "Do you own a dog?"

Responses of 40 Cat Owners to the Question "Do You Own a Dog?"

[= 2 cat owners]

Yes 

No 

(A) 6
(B) 12
(C) 14
(D) 28

20. Consider the following bar graph. Exactly how many students passed the chemistry course (that is, achieved a grade of D or better)?

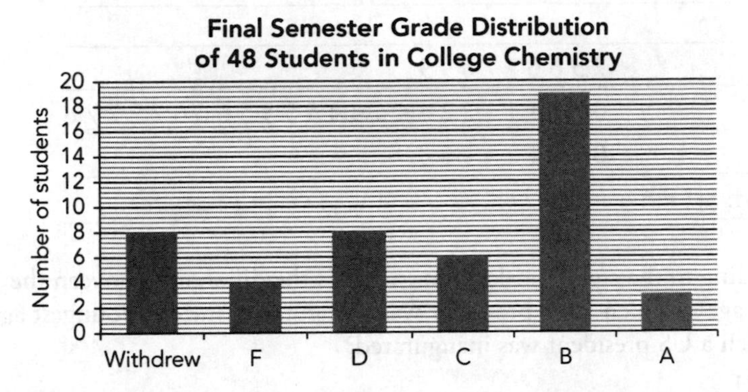

(A) 32
(B) 34
(C) 36
(D) 38

21. Consider the following double line graph. What was the lowest monthly profit (revenues – expenses)?

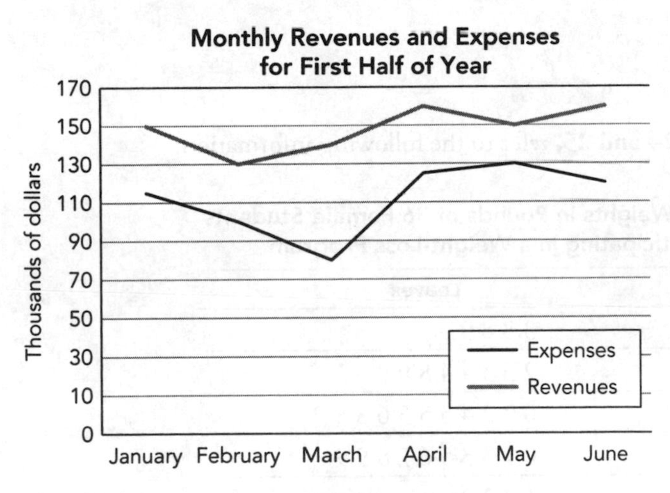

(A) $40,000
(B) $30,000
(C) $20,000
(D) $10,000

For questions 22 and 23, refer to the following information:

Age in Years of 44 US Presidents at Inauguration

Stem	Leaves
4	2 3 6 6 7 7 8 9 9
5	0 0 1 1 1 1 2 2 4 4 4 4 4 5 5 5 5 6 6 6 7 7 7 7 8
6	0 1 1 1 2 4 4 5 8 9
Legend: 5|7 = 57	

22. According to the stem-and-leaf plot, what is the difference between the oldest age at which a US president was inaugurated and the youngest age at which a US president was inaugurated?

(A) 11
(B) 18
(C) 20
(D) 27

23. According to the stem-and-leaf plot, how many of the 44 US presidents were 64 or older at inauguration?

(A) 10
(B) 5
(C) 4
(D) 3

For questions 24 and 25, refer to the following information:

Starting Weights in Pounds of 36 Female Students Participating in a Weight-Loss Program

Stem	Leaves
12	3 8 8 9
13	2 3 3 4 4 8 9
14	0 2 2 4 5 5 5 6 8 8 9
15	0 1 3 3 5 6 7 8 9 9
16	1 2 2 9
Legend: 12|3 = 123	

24. According to the stem-and-leaf plot, what starting weight occurs most often?

(A) 128
(B) 134
(C) 145
(D) 162

25. According to the stem-and-leaf plot, approximately what percent of the 36 female participants have a starting weight of at least 145 pounds?

(A) 33%
(B) 44%
(C) 50%
(D) 58%

For questions 26 and 27, refer to the information shown here:

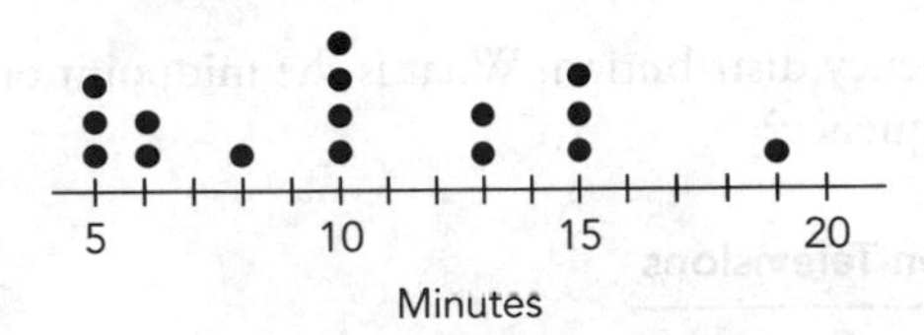

26. According to the dot plot, how many customers waited less than 15 minutes in line?

(A) 10
(B) 12
(C) 13
(D) 15

27. According to the dot plot, what is the greatest amount of time in minutes waited in line by a customer?

(A) 15
(B) 17
(C) 19
(D) 20

28. Consider the following frequency distribution. What is the class width of the distribution?

Prices of Seventy 46" Flat-Screen Televisions

Price (in US Dollars)	Frequency
500.00–799.99	15
800.00–1,099.99	25
1,100.00–1,399.99	13
1,400.00–1,699.99	9
1,700.00–1,999.99	8

(A) 99.99
(B) 299.99
(C) 1,399.99
(D) 1,499.99

29. Consider the following frequency distribution. What is the midpoint of the class with the highest frequency?

Prices of Fifty 46" Flat-Screen Televisions

Price (in US Dollars)	Frequency
500.00–799.99	10
800.00–1,099.99	17
1,100.00–1,399.99	11
1,400.00–1,699.99	8
1,700.00–1,999.99	4

(A) 649.995
(B) 949.995
(C) 1,249.995
(D) 1,549.995

For questions 30 and 31, refer to the following graph:

Weekly Soft Drink Consumption
of 120 High School Students

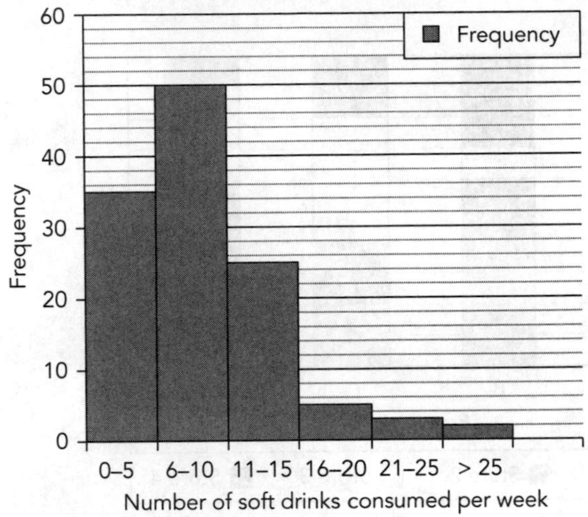

30. Based on the frequency histogram shown, about how many students consume fewer than 16 soft drinks per week?

(A) 10
(B) 50
(C) 85
(D) 110

31. Describe the shape of the distribution shown.

(A) skewed left
(B) negatively skewed
(C) positively skewed
(D) not skewed

32. Four stores (Store 1, Store 2, Store 3, and Store 4) jointly sold 100 units of each of four different products, A, B, C, and D. Based on the stacked bar chart shown, which store sold 25% of the 400 total units sold by the four stores?

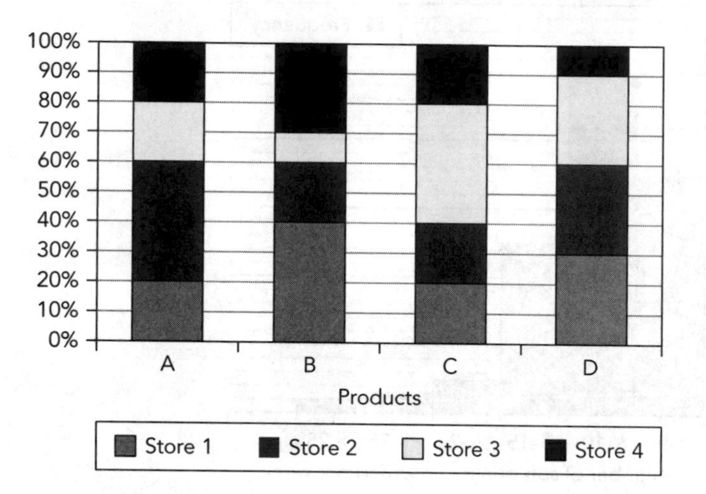

(A) Store 1
(B) Store 2
(C) Store 3
(D) Store 4

CHAPTER 3

Measures of Central Tendency

33. Find the mean of the following data:

15, 15, 33, 17, 30, 30, 20, 60, 45, 15

(A) 15
(B) 25
(C) 28
(D) 45

34. Find the mean of the following data:

−4, 25, −4, 11, 19, 4

(A) −4
(B) 7.5
(C) 8.5
(D) 51

35. Find the mean of the following data:

4.7, 5.6, 2.5, 4.9, 7.3, 4.7, 5.6, 6.5

(A) 5.225
(B) 5.250
(C) 5.275
(D) 5.600

36. Find the mean of the following data:

0, 0, 0, 0, 0, 100, 100, 100, 100, 1400

(A) 0
(B) 50
(C) 180
(D) 360

37. Find the mean of the following data:

−15, −15, −33, −17, −30, −30, −20, −60, −45, −15

(A) −15
(B) −25
(C) −28
(D) −45

38. Find the median of the following data:

15, 15, 33, 17, 30, 30, 20, 60, 45, 15

(A) 15
(B) 25
(C) 28
(D) 45

39. Find the median of the following data:

−4, 25, −4, 11, 19

(A) −4
(B) 7.5
(C) 8.5
(D) 11

40. Find the median of the following data:

4.7, 5.6, 2.5, 4.9, 7.3, 4.7, 5.6, 6.5

(A) 5.225
(B) 5.250
(C) 5.275
(D) 5.600

41. Find the median of the following data:

0, 0, 0, 0, 0, 100, 100, 100, 100, 1400

(A) 0
(B) 50
(C) 180
(D) 360

42. Find the median of the following data:

−15, −15, −33, −17, −30, −30, −20, −60, −45, −15

(A) −15
(B) −25
(C) −28
(D) −45

43. Find the mode of the following data:
15, 15, 33, 17, 30, 30, 20, 60, 45, 15

(A) 15
(B) 25
(C) 28
(D) There is no mode.

44. Find the mode of the following data:

−4, 25, 25, −4, 11, 11, 19, 19, 8, 8

(A) −4
(B) 11
(C) 12
(D) There is no mode.

45. Find the mode of the following data:

4.7, 5.6, 2.5, 4.9, 7.3, 4.7, 5.6, 6.5

(A) 4.7
(B) 5.6
(C) 4.7 and 5.6
(D) There is no mode.

46. Find the mode of the following data:

0, 0, 0, 0, 0, 100, 100, 100, 100, 1400

(A) 0
(B) 100
(C) 0 and 100
(D) There is no mode.

47. Find the mode of the following data:

−15, −15, −33, −17, −30, −15, −30, −20, −20, −60, −45, −15

(A) −15
(B) −15 and −30
(C) −15, −20, and −30
(D) There is no mode.

48. A student's exam score average is a weighted mean of five exam scores. Find the student's exam score average, where the weighted mean is computed according to the following table.

Exam	Score	Weight
Exam 1	80	10%
Exam 2	70	15%
Exam 3	55	20%
Exam 4	90	25%
Exam 5	60	30%

(A) 70
(B) 71
(C) 72
(D) 73

49. In determining the numerical course grade for a freshman psychology class, the instructor calculates a weighted average as shown in the table. A student has scores of 78, 81, and 75 on the 3 unit exams, an average of 92 on weekly quizzes, and a final-exam score of 75. To the nearest tenth, what is this student's numerical course grade?

Determination of Course Grade	Percent
Average (mean) of 3 unit exams	50%
Average (mean) of weekly quizzes	10%
Final-exam score	40%

(A) 78.2
(B) 80.2
(C) 82.2
(D) 84.2

50. In the following data set, what additional number should be included in the data set to yield a mean of 10?

5, 8, 10, 12, ___

(A) 8
(B) 10
(C) 12
(D) 15

51. A student has the following scores on three 100-point tests: 77, 91, and 94. What score must the student earn on the fourth 100-point test to have an average (mean) of 90 for the four tests?

(A) 90
(B) 95
(C) 98
(D) 100

52. Consider the following line graph. To the nearest thousand dollars, determine the mean monthly revenues.

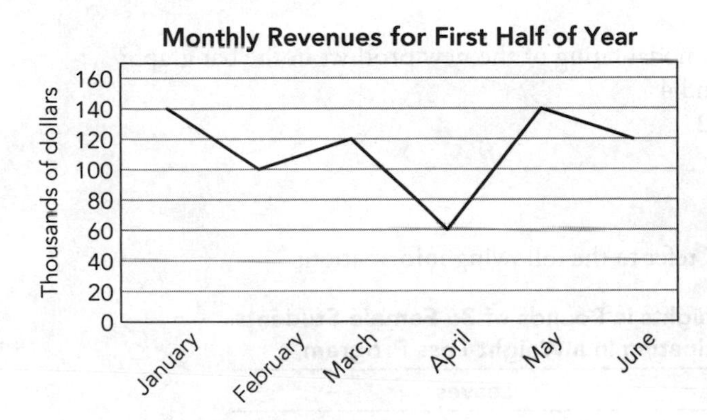

(A) $120,000
(B) $117,000
(C) $140,000
(D) $113,000

For questions 53 and 54, refer to the following graph:

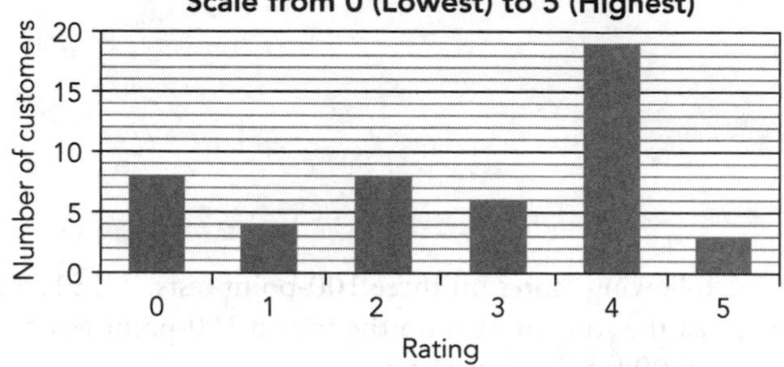

53. Considering the bar graph, determine the median rating of the new product.

(A) 2.5
(B) 3
(C) 3.5
(D) 4

54. What is the modal rating of the new product in the bar graph?

(A) 0, 2, and 4
(B) 0 and 2
(C) 4
(D) 19

Questions 55–57 refer to the following information:

Starting Weights in Pounds of 36 Female Students Participating in a Weight-Loss Program

Stem	Leaves
12	3 8 8 9
13	2 3 3 4 4 8 9
14	0 2 2 4 5 5 5 6 8 8 9
15	0 1 3 3 5 6 7 8 9 9
16	1 2 2 9
Legend: 12|3 = 123	

55. To the nearest tenth, what is the mean starting weight in pounds of the 36 participants?
(A) 140.0
(B) 145.0
(C) 145.5
(D) 145.8

56. To the nearest tenth, what is the median starting weight in pounds of the 36 participants?
(A) 140.0
(B) 145.0
(C) 145.5
(D) 145.8

57. What is the modal starting weight in pounds of the 36 participants?
(A) 128
(B) 145
(C) 162
(D) There is no mode.

Questions 58–61 refer to the following information:

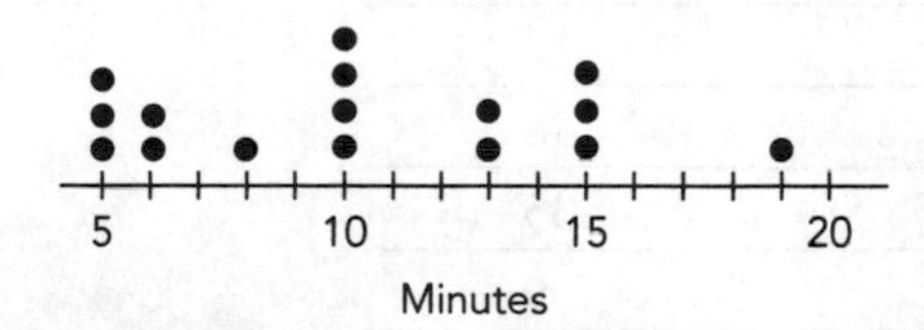

58. To the nearest tenth, what is the mean number of minutes the 16 customers waited in line?
(A) 10.0
(B) 10.3
(C) 10.5
(D) 11.0

59. According to the dot plot, to the nearest tenth, what is the median number of minutes the 16 customers waited in line?
(A) 10.0
(B) 10.3
(C) 10.5
(D) 12.5

60. According to the dot plot, what is the modal number of minutes the 16 customers waited in line?

(A) 5
(B) 10
(C) 15
(D) There is no mode.

61. Describe the skewness of the dot plot shown.

(A) strongly positively skewed
(B) slightly positively skewed
(C) strongly negatively skewed
(D) slightly negatively skewed

62. The table shown displays the classification of 240 movies according to genre. Which measure of central tendency is most appropriate for describing the data?

Genre	Number of Movies
Action/adventure	33
Drama	16
Comedy/romantic comedy	76
Musical	11
Mystery/crime	12
Horror/suspense/occult	48
Science fiction/fantasy	35
Other	9
Total	240

(A) mean
(B) median
(C) mode
(D) none of the above

63. Which measure of central tendency is most appropriate for describing the population data set associated with the histogram shown here?

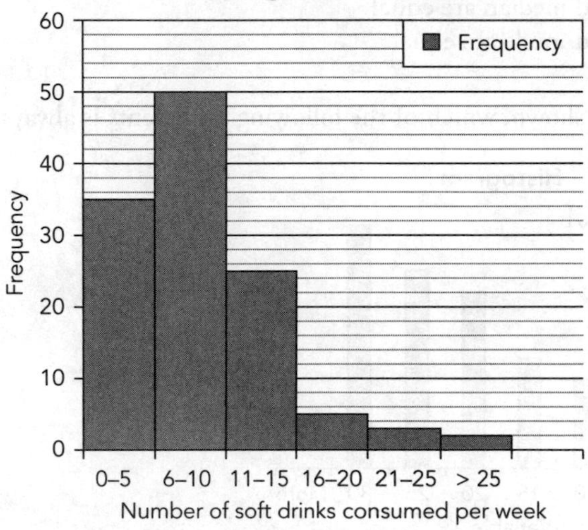

(A) mean
(B) median
(C) mode
(D) none of the above

64. Which measure of central tendency is most appropriate for describing the population data set associated with the graph shown?

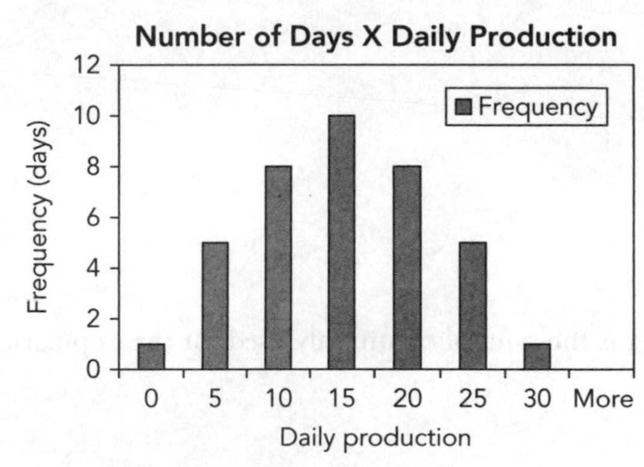

(A) mean
(B) median
(C) mode
(D) none of the above

65. Refer to graph for question 63. For this histogram, which of the following statements is always true?

(A) The mean is greater than the median.
(B) The median is greater than the mean.
(C) The mean and median are equal.
(D) The mean and mode are equal.

66. For the histogram shown, which of the following statements is always true?

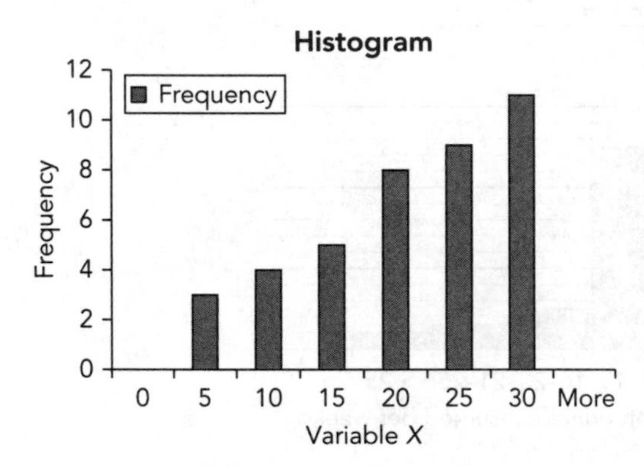

(A) The mean is greater than the median.
(B) The median is greater than the mean.
(C) The mean and median are equal.
(D) The mean and mode are equal.

67. Which measure(s) of central tendency is/are very sensitive to outliers?

I. mean
II. median
III. mode

(A) I only
(B) II only
(C) III only
(D) I and II only

68. Which of the following is the symbol commonly used for the population mean?

(A) σ
(B) σ^2
(C) μ
(D) $\overline{x}$

69. Which of the following is the symbol commonly used for the sample mean?

(A) σ
(B) σ^2
(C) μ
(D) $\overline{x}$

70. Consider the stem-and-leaf plot shown. Which of the following best describes the shape of this distribution?

Stem	**Leaves**
1	5 5 5 5
2	5 5 5 5 5 5 5
3	5 5 5 5 5 5 5 5 5 5 5 5 5 5
4	5 5 5 5 5 5 5
5	5 5 5 5
Legend: 1|5 = 15	

(A) skewed right
(B) skewed left
(C) symmetric—unimodal
(D) symmetric—multimodal

CHAPTER 4

Measures of Variability

71. Find the range of the following data:

15, 33, 30, 50, 0

(A) 15
(B) 20
(C) 17
(D) 50

72. Find the range of the following data:

−4, 25, −4, 11, 19, 4

(A) 8
(B) 15
(C) 21
(D) 29

73. Find the range of the following data:

−15, −15, −33, −17, −30, −30, −20, −60, −45, −15

(A) −75
(B) −45
(C) 45
(D) 75

74. Find the range of the data associated with the line graph shown.

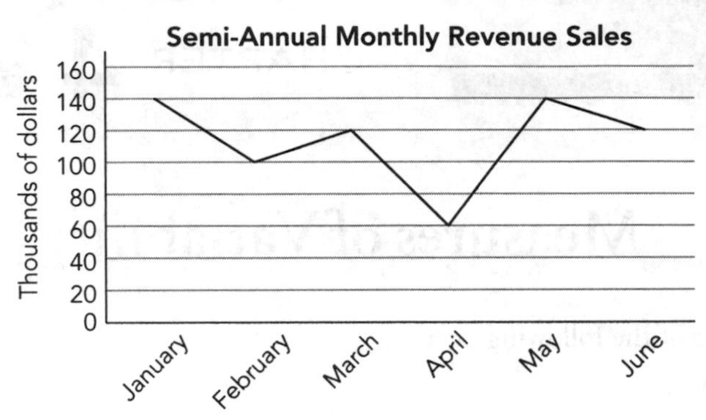

(A) \$20,000
(B) \$40,000
(C) \$60,000
(D) \$80,000

75. Find the range of the data associated with the stem-and-leaf plot shown.

Starting Weights in Pounds of 36 Female Students Participating in a Weight-Loss Program

Stem	Leaves
12	3 8 8 9
13	2 3 3 4 4 8 9
14	0 2 2 4 5 5 5 6 8 8 9
15	0 1 3 3 5 6 7 8 9 9
16	1 2 2 9
Legend: 12|3 = 123	

(A) 7
(B) 9
(C) 40
(D) 46

76. Find the range of the data associated with the dot plot shown.

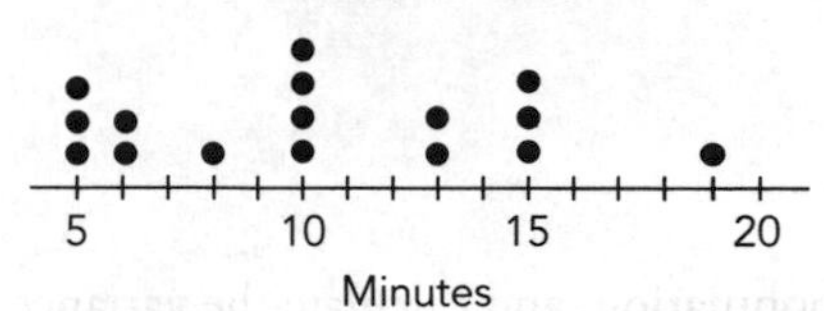

(A) 14
(B) 15
(C) 19
(D) 20

77. Both the variance and standard deviation are intended to provide a measure of the variability of a set of data values about which of the following?

(A) mean
(B) median
(C) mode
(D) range

78. Which of the following symbols commonly is used for the population variance?

(A) σ
(B) σ^2
(C) s
(D) s^2

79. Which of the following symbols commonly is used for the sample variance?

(A) σ
(B) σ^2
(C) s
(D) s^2

80. Which of the following symbols commonly is used for the population standard deviation?

(A) σ
(B) σ^2
(C) s
(D) s^2

81. Which of the following symbols commonly is used for the sample standard deviation?

(A) σ
(B) σ^2
(C) s
(D) s^2

82. Consider the following data as a population, and calculate the variance to three decimal places.

15, 33, 30, 50, 0

(A) 359.300
(B) 287.440
(C) 18.955
(D) 16.954

83. Consider the following data as a sample, and calculate the variance to three decimal places.

15, 33, 30, 50, 0

(A) 359.300
(B) 287.440
(C) 18.955
(D) 16.954

84. Consider the following data as a population, and calculate the standard deviation to two decimal places.

–4, 25, –4, 11, 19, 4

(A) 144.30
(B) 120.25
(C) 12.01
(D) 10.97

85. Consider the following data as a sample, and calculate the standard deviation to two decimal places.

–4, 25, –4, 11, 19, 4

(A) 144.30
(B) 120.25
(C) 12.01
(D) 10.97

86. Consider the following data as a population, and calculate the variance to three decimal places.

4.7, 5.6, 2.5, 4.9, 7.3, 4.7, 5.6, 6.5

(A) 1.337
(B) 1.429
(C) 1.787
(D) 2.042

87. Consider the following data as a sample, and calculate the variance to three decimal places.

4.7, 5.6, 2.5, 4.9, 7.3, 4.7, 5.6, 6.5

(A) 1.337
(B) 1.429
(C) 1.787
(D) 2.042

88. Consider the following data as a population, and calculate the standard deviation to three decimal places.

4.7, 5.6, 2.5, 4.9, 7.3, 4.7, 5.6, 6.5

(A) 1.337
(B) 1.429
(C) 1.787
(D) 2.042

89. Consider the following data as a sample, and calculate the standard deviation to three decimal places.

4.7, 5.6, 2.5, 4.9, 7.3, 4.7, 5.6, 6.5

(A) 1.337
(B) 1.429
(C) 1.787
(D) 2.042

90. Consider the following data as a population, and calculate the standard deviation to three decimal places.

−70, −60, −50, −50, −40, −30

(A) −12.910
(B) −14.142
(C) 14.142
(D) 12.910

91. Consider the following data as a sample, and calculate the standard deviation to three decimal places.

−70, −60, −50, −50, −40, −30

(A) −12.910
(B) −14.142
(C) 14.142
(D) 12.910

92. Using a graphing calculator, find the standard deviation of the sample data associated with the line graph shown. (Round your answer to the nearest dollar.)

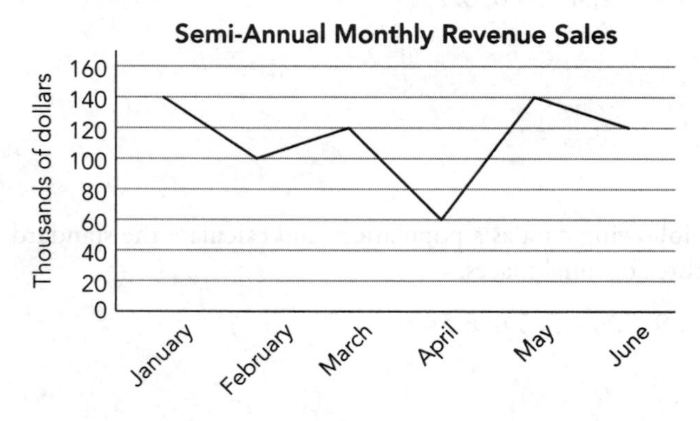

(A) \$11,333
(B) \$30,111
(C) \$27,487
(D) \$68,000

93. Using a graphing calculator, find the standard deviation of the sample data associated with the stem-and-leaf plot shown. (Round your answer, as needed.)

Stem	Leaves
10	3 8 8 9
11	2 3 3 4 4
12	0 2 2 8 8 9
Legend: 10|3 = 103	

(A) 66.60
(B) 62.16
(C) 8.16
(D) 7.88

94. Using a graphing calculator, find the standard deviation of the sample data associated with the dot plot shown. (Round your answer, as needed.)

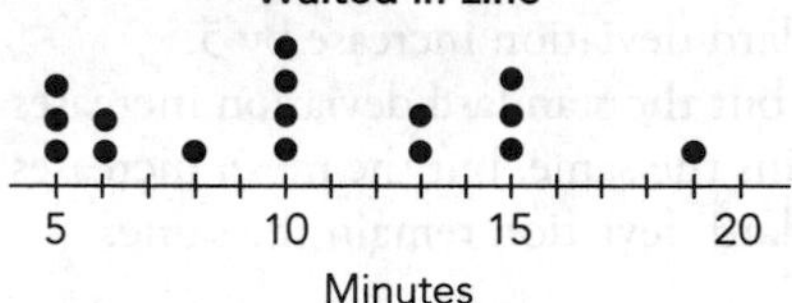

(A) 4.2
(B) 4.3
(C) 10.3
(D) 18.9

95. The summary statistics in the following table are based on five repetitions of the same experiment performed by four different groups of students: Group A, Group B, Group C, and Group D. The mean value of which group is most reliable?

Summary Statistics of Experiment Based on 5 Trials

Group	Mean	Standard Deviation
A	25 cm	5 cm
B	28 cm	7 cm
C	31 cm	6 cm
D	25 cm	9 cm

(A) Group A
(B) Group B
(C) Group C
(D) Group D

96. Consider the following data as a sample. Which of the following statements is true?

50, 50, 50, 50, 50, 50, 50, 50, 50, 50, 50, 50, 50, 50, 50

(A) The data are positively skewed.
(B) The mode is 15.
(C) The variance is zero.
(D) The range is 50.

97. Consider the following data as a sample. If each data value is increased by 5, what is the effect on the mean and standard deviation?

15, 15, 33, 17, 30, 30, 20, 60, 45, 15

(A) Both the mean and the standard deviation increase by 5.
(B) The mean remains the same, but the standard deviation increases by 5.
(C) The standard deviation remains the same, but the mean increases by 5.
(D) Both the mean and the standard deviation remain the same.

98. Which of the following measures is the best measure of location for describing skewed population data?

(A) mean
(B) median
(C) mode
(D) range

99. Which of the following measures is the best measure of location for describing sample data?

(A) mean
(B) median
(C) mode
(D) standard deviation

100. Consider the sample data displayed in the following dot plot. Find Pearson's coefficient of skewness for these data. (Round your answer, as needed.)

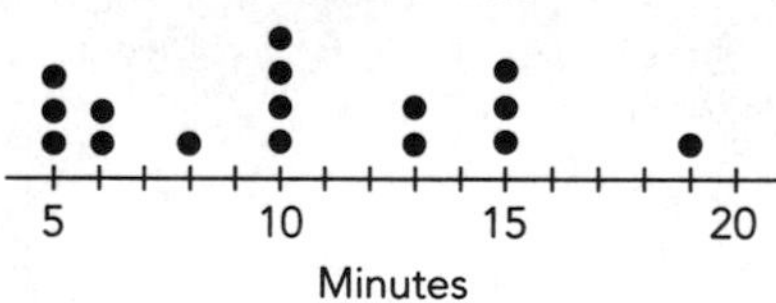

(A) −0.222
(B) −0.216
(C) 0.216
(D) 0.222

101. Consider the following data as a population, and find Pearson's coefficient of skewness for these data. (Round your answer, as needed.)

0, 0, 0, 0, 0, −100, −100, −100, −100, −1,400

(A) −0.953
(B) −0.904
(C) 0.904
(D) 0.953

102. Data from a population yield $\mu = 360$ and $\sigma = 4.5$. Find the coefficient of variation for these data.

(A) 0.0125%
(B) 1.25%
(C) 80%
(D) 8,000%

103. Data from a sample yield $\bar{x} = 82$ and $s = 16.4$. Find the coefficient of variation for these data.

(A) 0.2%
(B) 20%
(C) 5%
(D) 500%

104. The summary statistics in the following table are based on five repetitions of the same experiment performed by four different groups of students: Group A, Group B, Group C, and Group D. The data of which group has the least relative variability?

Summary Statistics of Experiment Based on 5 Trials

Group	Mean $\bar{x}$	Standard Deviation s
A	25 cm	5 cm
B	28 cm	7 cm
C	31 cm	6 cm
D	25 cm	9 cm

(A) Group A
(B) Group B
(C) Group C
(D) Group D

CHAPTER 5

Percentiles and the Five-Number Summary

105. Consider the following data and determine the 80th percentile.

15, 15, 33, 17, 30, 30, 20, 60, 45, 15

(A) 8
(B) 33
(C) 39
(D) 45

106. Consider the following data and determine the 50th percentile.

–4, 25, –4, 11, 19, 4

(A) –4
(B) 3.5
(C) 7.5
(D) 11

107. Consider the following data and determine the 70th percentile.

4.5, 5.6, 2.5, 4.5, 7.3, 4.8, 6.6, 6.6

(A) 5.6
(B) 6.1
(C) 6.6
(D) 5.2

108. Consider the following data and determine the 90th percentile.

0, 0, 0, 0, 0, 100, 100, 100, 100, 1400

(A) 90
(B) 100
(C) 750
(D) 1400

109. Consider the following data and determine the 35th percentile.

−15, −15, −33, −17, −30, −30, −20, −60, −45, −15

(A) −17
(B) −39
(C) −33
(D) −30

110. Consider the bar graph shown. Determine the 25th percentile.

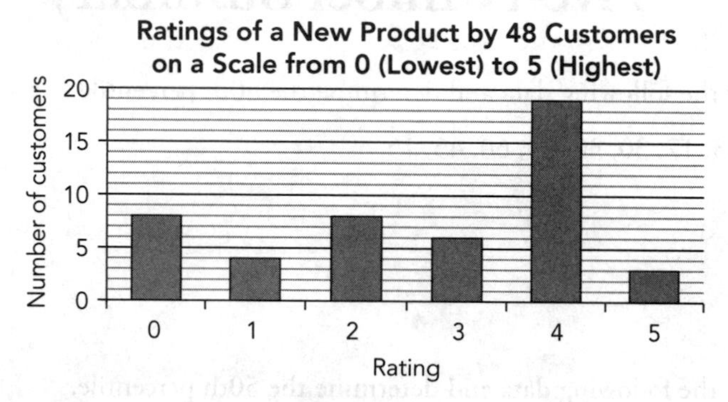

(A) 1
(B) 1.5
(C) 2
(D) 2.5

111. Consider the stem-and-leaf plot shown. Determine the 95th percentile.

Starting Weights in Pounds of 36 Female Students Participating in a Weight-Loss Program

Stem	Leaves
12	3 8 8 9
13	2 3 3 4 4 8 9
14	0 2 2 4 5 5 5 6 8 8 9
15	0 1 3 3 5 6 7 8 9 9
16	1 2 5 9

Legend: 12|3 = 123

(A) 162
(B) 165
(C) 167
(D) 169

112. Consider the dot plot shown. Determine the 37.5th percentile.

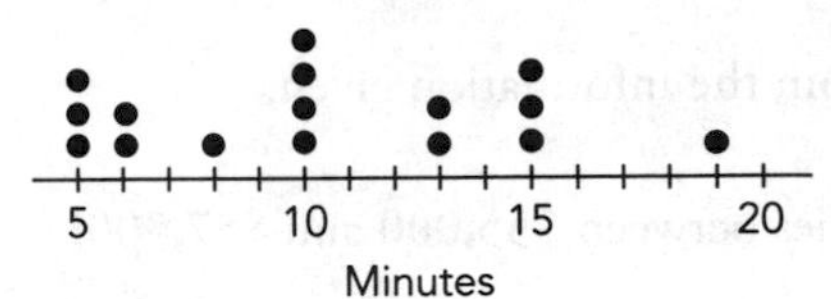

(A) 6
(B) 8
(C) 9
(D) 10

Questions 113–120 refer to the following information. In a population of 320 employees of a manufacturing plant, the mean salary is $55,000 with a standard deviation of $4,000. The median salary is $48,000, and the 60th percentile is $57,500.

113. Which of the following statements is always true?
(A) The salary data are skewed to the left.
(B) Fifty percent of the employees have salaries of at least $55,000.
(C) Sixty percent of the employees have salaries that are less than $57,500.
(D) Ten percent of the employees have salaries between $55,000 and $57,500.

114. How many employees have salaries that are greater than $57,500?
(A) 40
(B) 128
(C) 160
(D) 192

115. Which of the following salaries has a percentile rank of 50?
(A) $48,000
(B) $51,500
(C) $55,000
(D) $57,500

116. How many employees have salaries between $48,000 and $57,500?

(A) 10
(B) 32
(C) 50
(D) It cannot be determined from the information given.

117. How many employees have salaries between $55,000 and $57,500?

(A) 10
(B) 32
(C) 50
(D) It cannot be determined from the information given.

118. What is the *z*-score for a salary of $57,500?

(A) 0.625
(B) 0.600
(C) 2.375
(D) 1.600

119. What is the *z*-score for the median?

(A) −1.75
(B) −0.500
(C) 0.500
(D) 1.75

120. Which of the following salaries has a *z*-score of 1.25?

(A) $50,000
(B) $53,000
(C) $60,000
(D) $62,500

Questions 121 and 122 refer to the following information. The mean weight of a population of 4,000 middle-school girls is 80 pounds with a standard deviation of 14 pounds. The 15th percentile has a *z*-score of −1.5.

121. What weight (in pounds) is the 15th percentile?

(A) 21
(B) 59
(C) 61
(D) 101

122. What weight in pounds has a z-score of 0.5?

(A) 40
(B) 63
(C) 79.5
(D) 87

123. The information in the following table shows a student's scores on four exams in a college biology class along with the means and standard deviations of the scores for all the students in the class of 50 students. On which of the exams did the student perform best relative to the performance of the student's classmates?

	Exam 1	Exam 2	Exam 3	Exam 4
Student's grade	75	87	92	70
Class mean	65	88	86	60
Class standard deviation	5	2	4	10

(A) Exam 1
(B) Exam 2
(C) Exam 3
(D) Exam 4

For questions 124–127, refer to the following box plot:

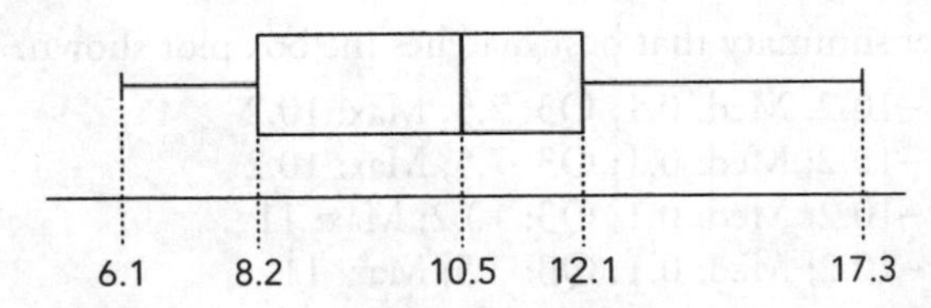

124. What is the median of the data set whose box plot is shown?

(A) 8.2
(B) 10.1
(C) 10.5
(D) 12.1

125. What is the range of the data set whose box plot is shown?

(A) 11.2
(B) 9.1
(C) 6.8
(D) 3.9

126. What is the IQR of the data set whose box plot is shown?

(A) 11.2
(B) 9.1
(C) 6.8
(D) 3.9

127. Which of the following statements is always true of the box plot shown?

(A) There are no outliers in the data.
(B) A value of 17.3 is a possible outlier.
(C) Any value between 6.1 and 8.2 is a possible outlier.
(D) Any value between 12.1 and 17.3 is a possible outlier.

For questions 128 and 129, refer to the following box plot.

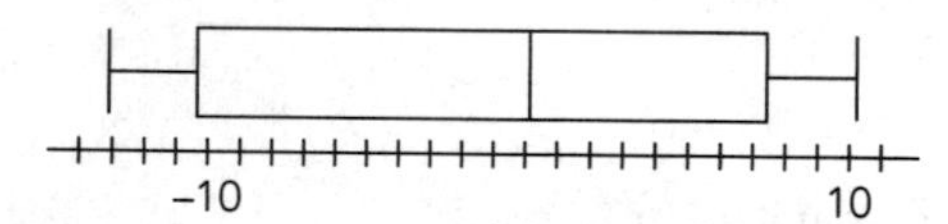

128. Choose the 5-number summary that best matches the box plot shown.

(A) Min: −11; Q1: −10.2; Med: 0.1; Q3: 7.5; Max: 10.2
(B) Min: −13; Q1: −10.2; Med: 0.1; Q3: 7.5; Max: 10.2
(C) Min: −13; Q1: −10.2; Med: 0.1; Q3: 10.2; Max: 11
(D) Min: −14; Q1: −10.2; Med: 0.1; Q3: 7.5; Max: 11

129. What is the approximate IQR of the data set whose box plot is shown?

(A) −2.7
(B) 17.7
(C) 2.7
(D) 23.2

130. Consider the bar graph shown and determine the IQR for the data.

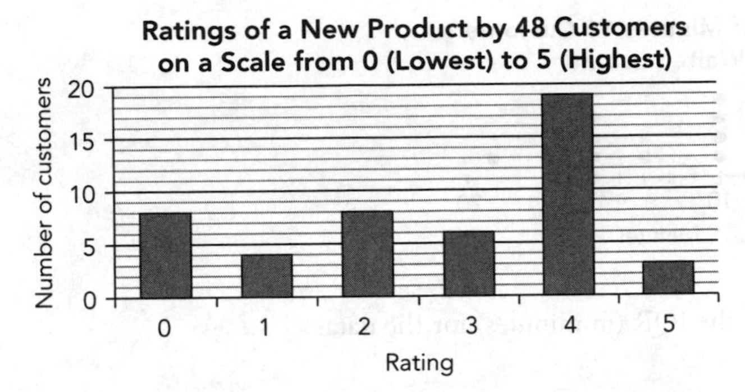

(A) 2
(B) 2.5
(C) 3
(D) 3.5

131. Consider the stem-and-leaf plot shown. Determine the IQR (in pounds) for the data.

Starting Weights in Pounds of 36 Female Students Participating in a Weight-Loss Program

Stem	Leaves
12	3 8 8 9
13	2 3 3 4 4 8 9
14	0 2 2 4 5 5 5 6 8 8 9
15	0 1 3 3 5 6 7 8 9 9
16	1 2 2 9
Legend: 12|3 = 123	

(A) 19
(B) 20
(C) 21.5
(D) 19.5

Questions 132 and 133 refer to the following figure.

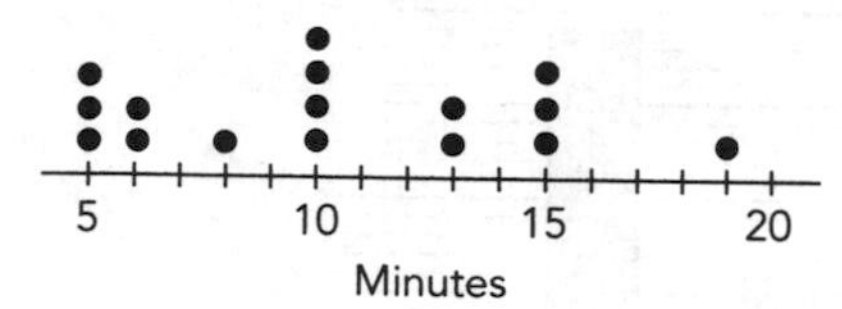

132. Determine the IQR (in minutes) for the data.

(A) 5
(B) 8
(C) 9
(D) 14

133. Which of the following statements is always true?

(A) There are no outliers in the data.
(B) A time of 19 minutes is a possible outlier.
(C) A time of 5 minutes is a possible outlier.
(D) Outliers cannot be determined for the dot plot shown.

134. Which of the following statements is always true about outliers?

(A) The maximum and minimum values of a data set are outliers.
(B) All outliers are the result of measurement errors.
(C) Outliers signal a concern that should be investigated.
(D) Data values that are more than one IQR below Q1 are outliers.

CHAPTER 6

Counting Techniques

135. How many different meal combinations consisting of one sandwich, one drink, and one type of chips are possible from a selection of 8 kinds of sandwiches, 5 drinks, and 7 types of chips?

(A) 20
(B) 35
(C) 40
(D) 280

136. A code for a locking briefcase consists of five digits that must be entered in a definite order. Each of the digits 0 through 9 may be used in the code, and repetition of digits is allowed. How many different five-digit codes are possible?

(A) 100,000
(B) 59,049
(C) 30,240
(D) 15,120

137. There are 30 separate candidates for three vice president positions at a university. Assuming all 30 candidates are qualified to be selected for any one of the three VP positions, how many different ways can the positions be filled?

(A) 27,000
(B) 24,360
(C) 4,060
(D) 60

138. Suppose a high school issues each student a student identification (ID) number that consists of two letters followed by six digits (0 through 9). Assuming repetition of letters and digits is not allowed, how many unique student ID numbers are possible?

(A) $26^2 \cdot 9^6$

(B) $\binom{26}{2}\binom{10}{6}$

(C) $26^2 \cdot 10^6$

(D) $26 \cdot 25 \cdot 10 \cdot 9 \cdot 8 \cdot 7 \cdot 6 \cdot 5$

139. In a certain state, the alphanumeric code for a car license plate consists of three letters followed by three digits (0 through 9). For that state, how many different car license plate alphanumeric codes are possible if repetition of digits and letters is allowed?

(A) 36^6

(B) $26^3 \cdot 10^3$

(C) $26^3 \cdot 9^3$

(D) $\binom{26}{3}\binom{10}{3}$

140. For the area code 936, how many seven-digit telephone numbers that begin with 564- are possible if repetition of digits is allowed?

(A) 10^4

(B) $3 \cdot 10^4$

(C) $10 \cdot 9 \cdot 8 \cdot 7$

(D) 10^7

141. Suppose a code is dialed by means of three disks, each of which is stamped with 15 letters. How many three-letter codes are possible using the three disks?

(A) ${}_{15}C_3$

(B) 15^3

(C) $15 \cdot 14 \cdot 13$

(D) $15 \cdot 3$

142. How many three-digit codes are possible using only the digits 1 and 0 if repetition of digits is allowed?

(A) 2
(B) 4
(C) 6
(D) 8

143. In how many different ways is it possible to seat three people in three chairs?

(A) 1
(B) 6
(C) 9
(D) 27

144. How many different ways can you arrange the five letters in the word *chair* if you use all five letters each time?

(A) ${}_5P_5$
(B) ${}_5C_5$
(C) 5^5
(D) 5^2

145. Four people are to be seated in four identical chairs placed in a circle. How many different arrangements of the four people (relative to one another) in the four chairs are possible?

(A) 4^4
(B) ${}_4C_4$
(C) ${}_4P_4$
(D) ${}_3P_3$

146. Eight of 10 people are selected to be seated in 8 chairs placed in a straight row. How many different arrangements of the 10 people in the 8 chairs are possible?

(A) 10^8
(B) $10 \cdot 8$
(C) ${}_{10}P_8$
(D) ${}_{10}C_8$

147. How many different 10-letter arrangements of the 10 letters in the word *statistics* are possible?

(A) $\frac{10!}{2!}$

(B) $\frac{10!}{2!8!}$

(C) $\frac{10!}{3!3!2!}$

(D) $\frac{10!}{3!2!}$

148. How many different ways can a club of 25 members fill the positions of president, vice president, and secretary from its membership if no person holds more than one office and all members are eligible for any one of the three positions?

(A) 75
(B) 2,300
(C) 13,800
(D) 15,625

149. How many six-digit (0 through 9) passwords are possible if repetition of digits is not allowed?

(A) 10^6
(B) ${}_{10}P_6$
(C) ${}_{10}C_6$
(D) ${}_{10}P_4$

150. How many different ways can a club of 20 members fill the positions on a 3-member officer-nominating committee from its membership if all members are eligible to serve on the committee?

(A) 60
(B) 1,140
(C) 6,840
(D) 8,000

151. How many 5-card hands are possible from a standard deck of 52 playing cards if the cards are drawn without replacement?

(A) 52^5
(B) ${}_{52}P_5$
(C) ${}_{52}C_5$
(D) $52 \cdot 5$

152. How many combinations of 3 pizza toppings can be created from a choice of 12 toppings if each topping can be used only once on a pizza?

(A) 48
(B) 220
(C) 1,320
(D) 1,728

153. How many combinations of 7 out of 15 patients with the same illness can be randomly selected to receive an experimental drug?

(A) $\frac{15!}{7!8!}$

(B) $\frac{15!}{8!}$

(C) $\frac{15!}{7!}$

(D) 15^3

154. How many different wardrobes can be created by selecting 5 out of 8 different shirts, 4 out of 10 different pairs of slacks, and 3 out of 6 different ties?

(A) $5 \cdot 8 + 4 \cdot 10 + 3 \cdot 6$
(B) $5^8 \cdot 4^{10} \cdot 3^6$
(C) ${}_8C_5 \cdot {}_{10}C_4 \cdot {}_6C_3$
(D) ${}_8P_5 \cdot {}_{10}P_4 \cdot {}_6P_3$

CHAPTER 7

Basic Probability Concepts

155. Two cards are drawn at random, successively without replacement, from a well-shuffled standard deck of 52 playing cards. Determine the number of outcomes in the sample space of this experiment.

(A) 104
(B) 2,601
(C) 2,652
(D) 2,704

156. A fair coin is flipped six times, and the up face on each coin is observed. Determine the number of outcomes in the sample space of this experiment.

(A) 2
(B) 12
(C) 36
(D) 64

157. Find the probability that a number greater than 4 appears on the up face in a single toss of a fair six-sided die.

(A) $\frac{1}{6}$

(B) $\frac{1}{3}$

(C) $\frac{1}{2}$

(D) $\frac{2}{3}$

158. Find the probability that a prime or an even number appears on the up face in a single toss of a fair six-sided die.

(A) $\frac{1}{3}$

(B) $\frac{2}{3}$

(C) $\frac{5}{6}$

(D) 1

159. Find the probability that a seven appears on the up face in a single toss of a fair six-sided die.

(A) 0

(B) $\frac{1}{6}$

(C) 1

(D) $\frac{7}{6}$

160. Find the probability that at least one head appears when two fair coins are flipped and the up face on each is observed.

(A) $\frac{1}{4}$

(B) $\frac{1}{2}$

(C) $\frac{3}{4}$

(D) 1

161. Find the probability that the number of dots on the up faces sum to 7 when a pair of fair six-sided dice is tossed and the up face on each is observed.

(A) $\frac{1}{12}$

(B) $\frac{1}{6}$

(C) $\frac{1}{4}$

(D) $\frac{7}{36}$

162. Find the probability of drawing a face card when a single card is drawn at random from a well-shuffled standard deck of 52 playing cards.

(A) $\frac{1}{12}$

(B) $\frac{3}{13}$

(C) $\frac{4}{13}$

(D) $\frac{1}{4}$

163. Find the probability of drawing a diamond or a 10 when a single card is drawn at random from a well-shuffled standard deck of 52 playing cards.

(A) $\frac{1}{13}$

(B) $\frac{7}{26}$

(C) $\frac{4}{13}$

(D) $\frac{6}{13}$

164. Find the probability of drawing a black or red marble when a single marble is drawn at random from an urn containing 7 black marbles, 6 green marbles, and 10 red marbles. (Assume the marbles are identical except for color.)

(A) $\frac{7}{23}$

(B) $\frac{10}{23}$

(C) $\frac{13}{23}$

(D) $\frac{17}{23}$

165. A quiz consists of five multiple-choice questions, each of which has four possible answer choices (A, B, C, and D), one of which is correct. Suppose that an unprepared student does not read the questions but simply makes a random guess for each question. What is the probability that the student will get all five questions wrong?

(A) 0

(B) $\frac{1}{1{,}024}$

(C) $\frac{243}{1{,}024}$

(D) $5\left(1-\frac{1}{4}\right)$

166. A box contains 30 identical-looking items, of which 3 are defective. If one item is selected at random, what is the probability that the item is defective?

(A) $\frac{1}{30}$

(B) $\frac{1}{11}$

(C) $\frac{1}{10}$

(D) $\frac{9}{10}$

167. Find the complement of the following event: A number greater than four appears on the up face in a single toss of a fair six-sided die.

(A) $\frac{5}{6}$

(B) $\frac{1}{3}$

(C) $\frac{1}{2}$

(D) $\frac{2}{3}$

168. Find the complement of the following event: A prime or an even number appears on the up face in a single toss of a fair six-sided die.

(A) $\frac{1}{3}$

(B) $\frac{2}{3}$

(C) $\frac{1}{6}$

(D) 0

169. Find the complement of the following event: drawing a black or red marble when a single marble is drawn at random from an urn containing 7 black marbles, 6 green marbles, and 10 red marbles. (Assume the marbles are identical except for color.)

(A) $\frac{16}{23}$

(B) $\frac{10}{23}$

(C) $\frac{6}{23}$

(D) $\frac{3}{23}$

170. Find the complement of the following event: At least one head appears when two fair coins are flipped and the up face on each is observed.

(A) $\frac{1}{4}$

(B) $\frac{1}{2}$

(C) $\frac{3}{4}$

(D) 0

171. Find the complement of the following event: The number of dots on the up faces sum to 7 when a pair of fair dice is tossed and the up face on each is observed.

(A) $\frac{11}{12}$

(B) $\frac{5}{6}$

(C) $\frac{3}{4}$

(D) $\frac{29}{36}$

172. A single marble is drawn at random from an urn containing 2 white marbles, 7 black marbles, 6 green marbles, and 10 red marbles. (Assume the marbles are identical except for color.) What is the probability that the marble drawn is NOT red?

(A) $\frac{2}{5}$

(B) $\frac{3}{5}$

(C) $\frac{1}{3}$

(D) $\frac{2}{3}$

173. A quiz consists of five multiple-choice questions, each of which has four possible answer choices (A, B, C, and D), one of which is correct. Suppose that an unprepared student does not read the questions but simply makes a random guess for each question. What is the probability that the student will guess correctly on at least one question?

(A) $\frac{1}{1,024}$

(B) $1-\frac{1}{1,024}$

(C) $1-\frac{243}{1,024}$

(D) $5\cdot\frac{1}{4}$

174. A box contains 30 identical-looking items, of which 3 are defective. If one item is selected at random, what is the probability that the item is NOT defective?

(A) $\frac{1}{10}$

(B) $\frac{29}{30}$

(C) $\frac{32}{33}$

(D) $\frac{9}{10}$

175. Of the following events X and Y, which pair of events is mutually exclusive?

(A) X = drawing a red card on one draw from a standard deck of 52 playing cards; Y = drawing a jack on one draw from a standard deck of 52 playing cards
(B) X = rolling a three on one toss of a fair six-sided die; Y = rolling a number greater than four on one toss of a fair six-sided die
(C) X = rolling an odd number on one toss of a fair six-sided die; Y = rolling a number less than three on one toss of a fair six-sided die
(D) X = drawing a face card on one draw from a standard deck of 52 playing cards; Y = drawing a king on one draw from a standard deck of 52 playing cards

176. Given $P(A) = 0.5$, $P(B) = 0.3$, and $P(A \text{ and } B) = 0.06$, calculate $P(A \text{ or } B)$.

(A) 0.15
(B) 0.94
(C) 0.80
(D) 0.74

177. Given $P(A) = 0.4$, $P(B) = 0.1$, and $P(A \cap B) = 0.05$, calculate $P(A \cup B)$.

(A) 0.04
(B) 0.45
(C) 0.95
(D) 0.50

178. Given $P(A) = 0.65$, $P(B) = 0.22$, and $P(A \cap B) = 0.08$, calculate $P(A \cup B)$.

(A) 0.87
(B) 0.92
(C) 0.79
(D) 0.14

179. Given $P(A) = \frac{4}{52}$, $P(B) = \frac{13}{52}$, and $P(A \text{ and } B) = \frac{1}{52}$, calculate $P(A \text{ or } B)$.

(A) $\frac{1}{52}$

(B) $\frac{4}{13}$

(C) $\frac{17}{52}$

(D) $\frac{51}{52}$

180. Given $P(A) = \frac{3}{8}$, $P(B) = \frac{5}{8}$, and $P(A \cap B) = 0$, calculate the probability that at least A or B occurs.

(A) 0

(B) $\frac{3}{8}$

(C) $\frac{15}{64}$

(D) 1

181. What is the probability of rolling an odd number or a number less than three on one roll of a fair six-sided die?

(A) $\frac{1}{3}$

(B) $\frac{2}{3}$

(C) $\frac{5}{6}$

(D) 1

182. One card is randomly drawn from a well-shuffled standard deck of 52 playing cards. Find the probability that the card drawn is an ace or a spade or a diamond.

(A) $\frac{15}{26}$

(B) $\frac{7}{13}$

(C) $\frac{8}{13}$

(D) $\frac{27}{52}$

183. Three fair coins are flipped, and the up face on each is observed. Find the probability that at least two heads are observed or the number of heads observed is an odd number.

(A) $\frac{1}{4}$

(B) $\frac{1}{2}$

(C) $\frac{7}{8}$

(D) 1

184. Three fair coins are flipped, and the up face on each is observed. Find the probability that all heads or all tails are observed.

(A) $\frac{1}{3}$

(B) $\frac{2}{3}$

(C) $\frac{1}{4}$

(D) $\frac{3}{4}$

185. Given $P(A) = 0.4$, $P(B) = 0.5$, and $P(A \text{ or } B) = 0.8$, find $P(A \text{ and } B)$.

(A) 0.1
(B) 0.2
(C) 0.6
(D) 0.9

186. Given $P(A) = 0.35$, $P(B) = 0.63$, and $P(A \cap B) = 0.32$, find $P(B|A)$.

(A) $\frac{32}{63}$

(B) $\frac{32}{35}$

(C) $\frac{7}{20}$

(D) It cannot be determined from the information given.

187. In which of the following circumstances are the events X and Y independent?

(A) $P(X) = \frac{1}{2}$, $P(Y) = \frac{1}{2}$, $P(X \cap Y) = \frac{1}{4}$

(B) $P(X) = 0.35$, $P(Y) = 0.63$, $P(X \cap Y) = 0.32$

(C) $P(X) = \frac{1}{2}$, $P(Y) = \frac{21}{55}$, $P(X \cap Y) = \frac{3}{22}$

(D) $P(X) = 0.7$, $P(X|Y) = 0.9$

188. What is the probability of rolling an odd number and then a number less than three on two rolls of a fair six-sided die?

(A) $\frac{1}{6}$

(B) $\frac{1}{4}$

(C) $\frac{2}{3}$

(D) $\frac{5}{6}$

189. Two cards are randomly drawn, successively without replacement, from a well-shuffled standard deck of 52 playing cards. Find the probability that the first card drawn is an ace and the second card drawn is a face card.

(A) $\frac{1}{3}$

(B) $\frac{4}{13}$

(C) $\frac{11}{663}$

(D) $\frac{4}{221}$

190. Three fair coins are flipped, and the up face on each is observed. Find the probability of observing all heads.

(A) $\frac{1}{8}$

(B) $\frac{1}{6}$

(C) $\frac{1}{4}$

(D) $\frac{1}{3}$

191. Two cards are randomly drawn, successively with replacement, from a well-shuffled standard deck of 52 playing cards. Find the probability that the first card drawn is a face card and the second card drawn is a number greater than 7 but less than 10.

(A) $\frac{8}{221}$

(B) $\frac{6}{169}$

(C) $\frac{12}{169}$

(D) $\frac{5}{13}$

192. Suppose you draw two marbles, successively with replacement, from a box containing eight red marbles and six blue marbles. Find the probability of drawing a blue marble on the second draw, given that you drew a red marble on the first draw. (Assume the marbles are identical except for color.)

(A) $\frac{12}{49}$

(B) $\frac{3}{7}$

(C) $\frac{6}{13}$

(D) $\frac{24}{91}$

193. Only one of 100 cell phones in a box is defective. The cell phones are randomly selected and tested one at a time. If the first 30 cell phones tested are not defective, what is the probability that the next cell phone selected and tested is defective?

(A) $\frac{99}{100}$

(B) $\frac{1}{70}$

(C) $\frac{3}{7}$

(D) $\frac{7}{10}$

194. The probability diagram shown represents the incidence of Internet failure during weather in which snow might occur. What is the probability that it snows and an Internet failure occurs?

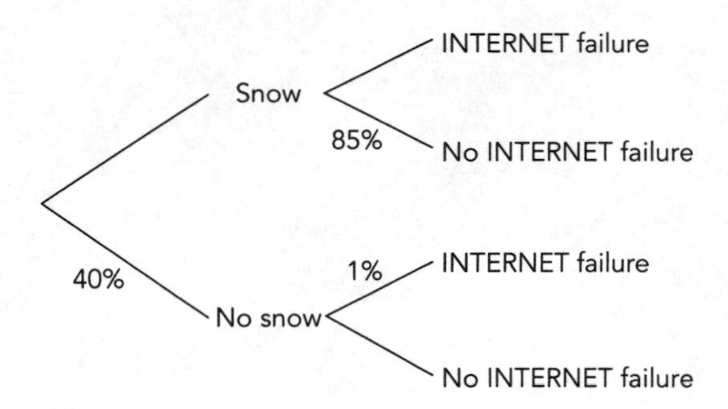

(A) 9%
(B) 0.4%
(C) 39.6%
(D) 51%

For questions 195 and 196, refer to the following table, which shows the residence status, by sex, of 250 seniors at a community college:

Residence Status of Senior Students (*n* = 250)

	On Campus	Off Campus
Female	52	86
Male	38	74

195. If one of the 250 students is randomly selected, what is the probability that the student resides on campus, given that the student selected is a male student? Express your answer as a decimal rounded to two places.

(A) 0.51
(B) 0.34
(C) 0.42
(D) 0.15

196. If one of the 250 students is randomly selected, what is the probability that the student is a female and resides off campus?

(A) 0.34
(B) 0.38
(C) 0.21
(D) 0.62

CHAPTER 8

Binomial Distribution

197. For which of the following experiments is the binomial distribution an appropriate model of the experiment's probability distribution?

(A) Toss a six-sided fair die 30 times, and record the up face of the die.

(B) Toss a six-sided fair die 100 times, and record the number of times the up face of the die shows five dots.

(C) Count the number of cars entering a car wash in a 30-minute period.

(D) Randomly draw 30 cards, successively without replacement, from a well-shuffled deck of 52 playing cards, and observe whether the card is a diamond.

198. For which of the following experiments is the binomial distribution NOT an appropriate model of the experiment's probability distribution?

(A) Toss two fair coins 10 times, and count the number of tosses in which two heads appear.

(B) For a quiz consisting of 10 multiple-choice questions, each with 4 possible answer choices (A, B, C, and D), one of which is correct, an unprepared student does not read the question but simply makes one random guess for each question. Count the number of correct guesses the student makes for the 5 questions.

(C) The probability that a randomly chosen remote control from a recent production lot will be defective is 0.15. Randomly select and test 10 remote controls from the production lot, and record the number of defective units.

(D) From an urn containing 6 white marbles and 9 black marbles, randomly draw 10 marbles, successively without replacement, and record the number of white marbles drawn.

199. Calculate μ for a binomial distribution with $n = 8$ repeated trials and probability of success $p = 0.2$. (Round your answer, as needed.)

(A) 1.13
(B) 1.28
(C) 1.6
(D) 6.4

200. Calculate σ^2 for a binomial distribution with $n = 8$ repeated trials and probability of success $p = 0.2$. (Round your answer, as needed.)

(A) 1.13
(B) 1.28
(C) 1.60
(D) 6.40

201. Calculate σ for a binomial distribution with $n = 8$ repeated trials and probability of success $p = 0.2$. (Round your answer, as needed.)

(A) 1.13
(B) 1.28
(C) 1.60
(D) 6.40

202. Calculate μ for a binomial distribution with $n = 15$ repeated trials and probability of success $p = 0.9$. (Round your answer, as needed.)

(A) 1.16
(B) 1.35
(C) 1.5
(D) 13.5

203. Calculate σ^2 for a binomial distribution with $n = 15$ repeated trials and probability of success $p = 0.9$. (Round your answer, as needed.)

(A) 1.16
(B) 1.35
(C) 1.5
(D) 13.5

204. Calculate σ for a binomial distribution with $n = 15$ repeated trials and probability of success $p = 0.9$. (Round your answer, as needed.)

(A) 1.16
(B) 1.35
(C) 1.5
(D) 13.5

205. Calculate μ for a binomial distribution with $n = 100$ repeated trials and probability of success $p = 0.5$. (Round your answer, as needed.)

(A) 100
(B) 50
(C) 25
(D) 5

206. Calculate σ^2 for a binomial distribution with $n = 100$ repeated trials and probability of success $p = 0.5$. (Round your answer, as needed.)

(A) 100
(B) 50
(C) 25
(D) 5

207. Calculate σ for a binomial distribution with $n = 100$ repeated trials and probability of success $p = 0.5$. (Round your answer, as needed.)

(A) 100
(B) 50
(C) 25
(D) 5

208. Calculate μ for a binomial distribution with $n = 300$ repeated trials and probability of success $p = 0.25$. (Round your answer, as needed.)

(A) 7.5
(B) 56.25
(C) 75
(D) 225

209. Calculate σ^2 for a binomial distribution with $n = 300$ repeated trials and probability of success $p = 0.25$. (Round your answer, as needed.)

(A) 7.5
(B) 56.25
(C) 75
(D) 225

210. Calculate σ for a binomial distribution with $n = 300$ repeated trials and probability of success $p = 0.25$. (Round your answer, as needed.)

(A) 7.5
(B) 56.25
(C) 75
(D) 225

211. A quality-control inspector randomly draws 20 light bulbs from a recent production lot and records the number of defective bulbs. Suppose it is known that the probability is 0.15 that a randomly chosen light bulb is defective. Calculate the expected number of defective bulbs.

(A) 3
(B) 5
(C) 10
(D) 17

For questions 212–214, use the formula for a binomial probability distribution to find the indicated probabilities for the binomial random variable *X*. (Round your answers, as needed.)

212. $P(X < 1)$ when $n = 5, p = 0.2$

(A) 0.328
(B) 0.737
(C) 0.032
(D) 0.672

213. $P(X \leq 1)$ when $n = 5, p = 0.2$

(A) 0.328
(B) 0.737
(C) 0.263
(D) 0.672

214. $P(X > 8)$ when $n = 10, p = 0.5$

(A) 0.989
(B) 0.055
(C) 0.011
(D) 0.945

For questions 215–225, use a graphing calculator to find the indicated probabilities for the binomial random variable *X*. (Round your answers, as needed.)

215. $P(X = 2)$ when $n = 3, p = 0.7$

(A) 0.216
(B) 0.441
(C) 0.343
(D) 0.657

216. $P(X \leq 2)$ when $n = 3, p = 0.7$

(A) 0.216
(B) 0.441
(C) 0.343
(D) 0.657

217. $P(X < 2)$ when $n = 3, p = 0.7$

(A) 0.216
(B) 0.441
(C) 0.343
(D) 0.657

218. $P(X > 2)$ when $n = 3, p = 0.7$

(A) 0.343
(B) 0.441
(C) 0.657
(D) 0.784

219. $P(X \geq 2)$ when $n = 3, p = 0.7$

(A) 0.343
(B) 0.441
(C) 0.657
(D) 0.784

220. $P(X = 5)$ when $n = 6, p = 0.4$

(A) 0.996
(B) 0.959
(C) 0.037
(D) 0.963

221. $P(X \geq 1)$ when $n = 10, p = 0.5$

(A) 0.055
(B) 0.001
(C) 0.989
(D) 0.999

222. $P(X$ is less than 1) when $n = 15, p = 0.2$

(A) 0.035
(B) 0.167
(C) 0.132
(D) 0.965

223. $P(X$ is not less than 8) when $n = 10, p = 0.9$

(A) 0.263
(B) 0.070
(C) 0.736
(D) 0.930

224. $P(X$ exceeds 8) when $n = 10, p = 0.5$

(A) 0.055
(B) 0.945
(C) 0.011
(D) 0.989

225. $P(X$ is less than or equal to 3) when $n = 12, p = 0.2$

(A) 0.205
(B) 0.795
(C) 0.558
(D) 0.236

For questions 226–231, use the binomial probabilities table in Appendix A to find the indicated probabilities for the binomial random variable X.

226. $P(X$ is exactly 17) when $n = 19, p = 0.8$

(A) 0.154
(B) 0.285
(C) 0.917
(D) 0.000

227. $P(X$ is at most 2) when $n = 11, p = 0.4$

(A) 0.120
(B) 0.031
(C) 0.089
(D) 0.880

228. $P(X$ is no less than 6) when $n = 9, p = 0.7$

(A) 0.267
(B) 0.156
(C) 0.463
(D) 0.730

229. $P(X$ is greater than 18) when $n = 20, p = 0.9$

(A) 0.285
(B) 0.323
(C) 0.392
(D) 0.677

230. $P(X$ is no more than 1) when $n = 15, p = 0.2$

(A) 0.035
(B) 0.132
(C) 0.167
(D) 0.833

231. $P(X$ is at least 15) when $n = 16, p = 0.7$

(A) 0.026
(B) 0.974
(C) 0.994
(D) 0.997

232. Toss a six-sided fair die five times, and record the number of dots on the up face of the die. What is the probability that an even number on the up face occurs exactly four times? (Round your answer, as needed.)

(A) 0.003
(B) 0.156
(C) 0.844
(D) 0.969

233. Flip a fair coin three times, and record the up face of the coin. What is the probability that heads shows on the up face no more than two times? (Round your answer, as needed.)

(A) 0.250
(B) 0.500
(C) 0.750
(D) 0.875

234. Suppose researchers determine that a new drug has a 40% chance of preventing a certain flu strain. If the drug is administered to 10 male subjects, what is the probability that the drug will be effective in preventing the flu strain for exactly eight of the male subjects? (Round your answer, as needed.)

(A) 0.012
(B) 0.998
(C) 0.011
(D) 0.989

235. Suppose researchers determine that the chances a pine tree in a particular forest is infected by the pine beetle is 10%. If five randomly selected pine trees in that forest are tested for infection by the pine beetle, what is the probability that exactly two are infected by the pine beetle? (Round your answer, as needed.)

(A) 0.927
(B) 0.008
(C) 0.073
(D) 0.991

236. A quiz consists of five multiple-choice questions, each with four possible answer choices (A, B, C, and D), one of which is correct. Suppose an unprepared student does not read the question but simply makes one random guess for each question. What is the probability that the student will get exactly five questions correct? (Round your answer, as needed.)

(A) 0.001
(B) 0.003
(C) 0.763
(D) 0.999

237. Toss a fair six-sided die 100 times, and record the up face each time. What is the probability that the up face of the die shows five dots at least 20 times? (Round your answer, as needed.)

(A) 0.152
(B) 0.220
(C) 0.780
(D) 0.848

238. A quiz consists of 10 multiple-choice questions, each with 5 possible answer choices (A, B, C, D, and E), one of which is correct. Suppose an unprepared student does not read the question but simply makes one random guess for each question. What is the probability that the student will get at least 3 questions correct? (Round your answer, as needed.)

(A) 0.322
(B) 0.001
(C) 0.678
(D) 0.879

CHAPTER 9

Normal Distribution

239. Consider the normal random variable X with mean $\mu = 75$ and standard deviation $\sigma = 5$. Find the probability that X assumes a value within one standard deviation of the mean.

(A) 0.0456
(B) 0.3174
(C) 0.6826
(D) 0.9974

240. Consider the normal random variable X with mean $\mu = 75$ and standard deviation $\sigma = 5$. Find the probability that X assumes a value more than one standard deviation from $\mu = 75$.

(A) 0.0456
(B) 0.3174
(C) 0.6826
(D) 0.9974

241. Consider the normal random variable X with mean $\mu = 300$ and standard deviation $\sigma = 20$. Find $P(260 < X < 340)$.

(A) 0.9544
(B) 0.9500
(C) 0.6826
(D) 0.9974

242. Consider the normal random variable X with mean $\mu = 300$ and standard deviation $\sigma = 20$. Find $1 - P(260 < X < 340)$.

(A) 0.0026
(B) 0.0456
(C) 0.6826
(D) 0.9974

243. Consider the normal random variable X with mean $\mu = 200$ and standard deviation $\sigma = 25$. Find $P(125 < X < 275)$.

(A) 0.3174
(B) 0.6826
(C) 0.9544
(D) 0.9974

244. Consider the normal random variable X with mean $\mu = 200$ and standard deviation $\sigma = 25$. Which of the following statements is always true?

(A) $P(X < 200)$ is less than $P(X > 200)$.
(B) $P(X < 200)$ is greater than $P(X > 200)$.
(C) $P(X = 200)$ equals 0.
(D) $P(X = 200)$ equals 0.5.

245. Consider the standard normal random variable Z with mean $\mu = 0$ and standard deviation $\sigma = 1$. Find $P(-1 < Z < 1)$.

(A) 0.3174
(B) 0.6826
(C) 0.9544
(D) 0.9974

246. Consider the standard normal random variable Z with mean $\mu = 0$ and standard deviation $\sigma = 1$. Find $P(-2 < Z < 2)$.

(A) 0.3174
(B) 0.6826
(C) 0.9544
(D) 0.9974

247. Consider the standard normal random variable Z with mean $\mu = 0$ and standard deviation $\sigma = 1$. Find $P(-3 < Z < 3)$.

(A) 0.3174
(B) 0.6826
(C) 0.9544
(D) 0.9974

248. Consider the standard normal random variable Z with mean $\mu = 0$ and standard deviation $\sigma = 1$. Find $P(Z \geq 0)$.

(A) 0.0000
(B) 0.2500
(C) 0.5000
(D) 0.9999

249. Consider the standard normal random variable Z with mean $\mu = 0$ and standard deviation $\sigma = 1$. Find $P(Z = 0)$.

(A) 0.0000
(B) 0.2500
(C) 0.5000
(D) 0.9999

For questions 250–255, use the standard normal distribution table in Appendix B to find the indicated probability.

250. $P(Z < 1.65)$

(A) 0.9505
(B) 0.4505
(C) 0.5495
(D) 0.5099

251. $P(Z < 0.08)$

(A) 0.5319
(B) 0.3188
(C) 0.7881
(D) 0.4681

252. $P(Z \leq -1.99)$

(A) 0.9767
(B) 0.4767
(C) 0.0233
(D) 0.5233

253. $P(Z \geq -1.04)$

(A) 0.1492
(B) 0.9192
(C) 0.0808
(D) 0.8508

254. $P(Z > 1.85)$

(A) 0.0322
(B) 0.9678
(C) 0.4678
(D) 0.0359

255. $P(-1.2 < Z < 2.4)$

(A) 0.1151
(B) 0.8767
(C) 0.1069
(D) 0.8933

For questions 256–268, use a graphing calculator to find the indicated probability. (Round your answer, as needed.)

256. $P(Z < -1.96)$

(A) 0.5250
(B) 0.4750
(C) 0.0250
(D) 0.9750

257. $P(Z \geq -1.96)$

(A) 0.5250
(B) 0.4750
(C) 0.0250
(D) 0.9750

258. $P(-1.96 < Z < 0)$

(A) 0.5250
(B) 0.4750
(C) 0.0250
(D) 0.9750

259. $P(0 < Z < 1.96)$

(A) 0.5250
(B) 0.4750
(C) 0.0250
(D) 0.9750

260. $P(-1.96 < Z < 1.96)$

(A) 0.0500
(B) 0.4750
(C) 0.9500
(D) 0.9750

261. $P(Z < -3)$
(A) 0.0013
(B) 0.4987
(C) 0.5013
(D) 0.9987

262. $P(Z < 3)$
(A) 0.0013
(B) 0.4987
(C) 0.5013
(D) 0.9987

263. $P(Z \leq 0)$
(A) 0.0000
(B) 0.2500
(C) 0.5000
(D) 1.0000

264. $P(Z > 2.33)$
(A) 0.9901
(B) 0.4901
(C) 0.0099
(D) 0.5099

265. $P(1.23 < Z \leq 1.87)$
(A) 0.0786
(B) 0.0307
(C) 0.1093
(D) 0.6400

266. $P(0 < Z < 1.65)$
(A) 0.4505
(B) 0.0495
(C) 0.9505
(D) 0.5495

267. $P(-1 < Z < 1)$
(A) 0.0456
(B) 0.3174
(C) 0.6827
(D) 0.9974

268. $P(-1.95 < Z < -1.28)$

(A) 0.0003
(B) 0.0747
(C) 0.6700
(D) 0.0256

For questions 269–273, use the standard normal distribution table in Appendix B to determine the percentile designation of the given *z*-score.

269. $z = -2.5$

(A) 48th
(B) 62nd
(C) 0.62th
(D) 0.48th

270. $z = 2.5$

(A) 62nd
(B) 99.46th
(C) 0.62th
(D) 99.38th

271. $z = -1.49$

(A) 6.81th
(B) 8.08th
(C) 0.68th
(D) 0.80th

272. $z = 1.64$

(A) 94.84th
(B) 94.95th
(C) 94.52th
(D) 5.05th

273. $z = 0$

(A) 0th
(B) 100th
(C) 25th
(D) 50th

For questions 274–278, use a graphing calculator to find the specified percentile z of the standard normal distribution. (Round your answers, as needed.)

274. 90th percentile

(A) $z = -1.28$
(B) $z = -1.34$
(C) $z = 1.34$
(D) $z = 1.28$

275. 20th percentile

(A) $z = -0.84$
(B) $z = -2.05$
(C) $z = 2.05$
(D) $z = 0.84$

276. 14th percentile

(A) $z = -2.99$
(B) $z = -1.08$
(C) $z = 1.08$
(D) $z = 2.99$

277. 60th percentile

(A) $z = -0.25$
(B) $z = -1.55$
(C) $z = 0.25$
(D) $z = 1.55$

278. 3rd percentile

(A) $z = -0.52$
(B) $z = -1.88$
(C) $z = 0.52$
(D) $z = 1.88$

For questions 279–283, use a graphing calculator to find the specified z_0 of the standard normal distribution. (Round your answers, as needed.)

279. $P(Z < z_0) = 0.8771$

(A) $z_0 = -1.16$
(B) $z_0 = 1.16$
(C) $z_0 = -2.38$
(D) $z_0 = 2.38$

280. $P(Z \le z_0) = 0.0124$

(A) $z_0 = -2.24$
(B) $z_0 = 2.24$
(C) $z_0 = 3.03$
(D) $z_0 = -3.03$

281. $P(Z \le z_0) = 0.9936$

(A) $z_0 = -1.29$
(B) $z_0 = -2.49$
(C) $z_0 = 1.29$
(D) $z_0 = 2.49$

282. $P(Z > z_0) = 0.975$

(A) $z_0 = -1.96$
(B) $z_0 = -1.30$
(C) $z_0 = 1.30$
(D) $z_0 = 1.96$

283. $P(Z \ge z_0) = 0.119$

(A) $z_0 = 2.37$
(B) $z_0 = -2.37$
(C) $z_0 = 1.18$
(D) $z_0 = -1.18$

For questions 284–297, use the standard normal distribution table in Appendix B, as needed, to find the indicated probability.

284. Consider the normal random variable X with mean $\mu = 60$ and standard deviation $\sigma = 10$. Find $P(X = 60)$.

(A) 0.0000
(B) 0.2500
(C) 0.5000
(D) 0.9999

285. Consider the normal random variable X with mean $\mu = 60$ and standard deviation $\sigma = 10$. Find $P(X \le 60)$.

(A) 0.0000
(B) 0.2500
(C) 0.5000
(D) 0.9999

286. Consider the normal random variable X with mean $\mu = 60$ and standard deviation $\sigma = 10$. Find $P(50 \leq X \leq 70)$.

(A) 0.3174
(B) 0.6826
(C) 0.9544
(D) 0.9974

287. Find the probability that a normal random variable X with mean $\mu = 25$ and standard deviation $\sigma = 5$ is less than 20.

(A) 0.1587
(B) 0.8413
(C) 0.6826
(D) 0.3174

288. Find the probability that a normal random variable X with mean $\mu = 300$ and standard deviation $\sigma = 50$ lies between 150 and 250.

(A) 0.0019
(B) 0.1587
(C) 0.8427
(D) 0.1573

289. Find the probability that a normal random variable X with mean $\mu = 30$ and standard deviation $\sigma = 6$ is at least 40.5.

(A) 0.4006
(B) 0.9599
(C) 0.0401
(D) 0.0004

290. Find the probability that a normal random variable X with mean $\mu = 120$ and standard deviation $\sigma = 25$ is no less than 100.

(A) 0.0788
(B) 0.0212
(C) 0.2119
(D) 0.7881

291. Find the probability that a normal random variable X with mean $\mu =$ 1,600 and standard deviation $\sigma = 200$ is at most 1,900.

(A) 0.9332
(B) 0.0668
(C) 0.0933
(D) 0.6680

292. Find the probability that a normal random variable X with mean $\mu =$ 1,600 and standard deviation $\sigma = 200$ is at least 1,900.

(A) 0.9332
(B) 0.0668
(C) 0.0933
(D) 0.6680

293. IQ scores for adults are normally distributed with mean $\mu = 100$ and standard deviation $\sigma = 15$. What percent of adults have IQ scores greater than 130?

(A) 0.9772
(B) 0.5987
(C) 0.0228
(D) 0.4013

294. Suppose that scores on a national exam are normally distributed with mean $\mu = 500$ and standard deviation $\sigma = 100$. Find the probability that a randomly selected test taker's score on the exam will be at least 750.

(A) 0.0062
(B) 0.9938
(C) 0.5987
(D) 0.4013

295. A local deli chain makes a signature sandwich. Suppose the measures of fat grams in the signature sandwiches are normally distributed with mean $\mu = 16$ grams and standard deviation $\sigma = 1.5$ grams. Find the probability that a randomly selected signature sandwich will have no more than 19 fat grams.

(A) 0.0228
(B) 0.9772
(C) 0.0977
(D) 0.2275

296. A plumbing company has found that the length of time, in minutes, for installing a bathtub is normally distributed with mean $\mu = 160$ minutes and standard deviation $\sigma = 25$ minutes. What percent of the bathtubs does the company install within 189 minutes?

(A) 0.8473
(B) 0.1523
(C) 0.8770
(D) 0.1230

297. The weights of newborn baby girls born at a local hospital are normally distributed with mean $\mu = 100$ ounces and standard deviation $\sigma = 8$ ounces. If a newborn baby girl born at the hospital is randomly selected, what is the probability that the child's weight will be at least 88 ounces?

(A) 0.0668
(B) 0.1479
(C) 0.4332
(D) 0.9332

For questions 298–301, use a graphing calculator to determine the probability. (Round your answers, as needed.)

298. The weights of newborn baby boys born at a local hospital are normally distributed with mean $\mu = 118$ ounces and standard deviation $\sigma = 9.5$ ounces. What percent of the baby boys born at the hospital weigh between 106.6 and 129.4 ounces?

(A) 76.99%
(B) 23.01%
(C) 38.49%
(D) 61.51%

299. Suppose the caloric content of a 12-ounce can of a certain diet drink is normally distributed with mean $\mu = 5$ calories and standard deviation $\sigma =$ 0.5 calories. What is the probability that a randomly selected 12-ounce can of the diet drink contains no more than 6 calories?

(A) 0.0228
(B) 0.9772
(C) 0.8643
(D) 0.1357

300. A coffee machine dispenses a mean of 8 ounces per cup. The machine's output is normally distributed with a standard deviation of 1 ounce. If the coffee cups filled by the machine hold 9.5 ounces, find the probability that the machine will overfill a cup.

(A) 0.9332
(B) 0.1440
(C) 0.0668
(D) 0.8560

301. The lifetime for a certain brand of incandescent light bulbs is normally distributed with a mean lifetime of 480 hours and standard deviation of 40 hours. What percent of these light bulbs last at least 500 hours?

(A) 69.15%
(B) 0.3085%
(C) 30.85%
(D) 0.6915%

302. Given that IQ scores are normally distributed with mean $\mu = 100$ and standard deviation $\sigma = 15$, find the 90th percentile of the distribution of IQ scores.

(A) 119.2
(B) 80.8
(C) 90.0
(D) 115.0

303. Given that IQ scores are normally distributed with mean $\mu = 100$ and standard deviation $\sigma = 15$, find the 20th percentile of the distribution of IQ scores.

(A) 112.6
(B) 87.4
(C) 20.0
(D) 84.0

304. Given that IQ scores are normally distributed with mean $\mu = 100$ and standard deviation $\sigma = 15$, find the 50th percentile of the distribution of IQ scores.

(A) 100
(B) 50
(C) 150
(D) 120

305. Consider the normal random variable X with mean $\mu = 500$ and standard deviation $\sigma = 100$. Find x_0 corresponding to $P(X < x_0) = 0.8771$.

(A) 116
(B) 877
(C) 588
(D) 616

306. Consider the normal random variable X with mean $\mu = 500$ and standard deviation $\sigma = 100$. Find x_0 corresponding to $P(X \le x_0) = 0.0124$.

(A) 276
(B) 512
(C) 224
(D) 124

307. Consider the normal random variable X with mean $\mu = 500$ and standard deviation $\sigma = 100$. Find x_0 corresponding to $P(X \le x_0) = 0.9936$.

(A) 249
(B) 599
(C) 994
(D) 749

308. Consider the normal random variable X with mean $\mu = 500$ and standard deviation $\sigma = 100$. Find x_0 corresponding to $P(X < x_0) = 0.025$.

(A) 205
(B) 250
(C) 304
(D) 498

309. Consider the normal random variable X with mean $\mu = 500$ and standard deviation $\sigma = 100$. Find x_0 corresponding to $P(X \le x_0) = 0.881$.

(A) 118
(B) 382
(C) 881
(D) 618

310. Suppose that scores on a national exam are normally distributed with mean $\mu = 500$ and standard deviation $\sigma = 100$. Ninety percent of test takers score less than what score on the exam?

(A) 450
(B) 372
(C) 590
(D) 628

311. A local deli chain makes a signature sandwich. Suppose the number of fat grams in the signature sandwiches is a normally distributed random variable with mean $\mu = 16$ grams and standard deviation $\sigma = 1.5$ grams. Find an amount of fat grams such that 50% of the sandwiches have no more than that many fat grams.

(A) 8
(B) 16
(C) 20
(D) 1.5

312. A plumbing company has found that the length of time, in minutes, for installing a bathtub is normally distributed with mean $\mu = 160$ minutes and standard deviation $\sigma = 25$ minutes. Eighty percent of the bathtubs are installed by the company within how many minutes?

(A) 181
(B) 128
(C) 80
(D) 84

313. The weights of newborn baby girls born at a local hospital are normally distributed with mean $\mu = 100$ ounces and standard deviation $\sigma = 8$ ounces. Find the weight (to the nearest ounce) for a newborn baby girl born at the hospital for which 25% of the newborn baby girls weigh less.

(A) 95
(B) 25
(C) 75
(D) 105

314. The lifetime for a certain brand of incandescent light bulbs is normally distributed with a mean lifetime of 480 hours and standard deviation of 40 hours. Find a lifetime (in hours) for this brand of incandescent light bulbs for which the probability is 0.9505 that a randomly selected incandescent light bulb of this brand will last at least that long.

(A) 480
(B) 482
(C) 546
(D) 414

CHAPTER 10

Estimation

315. A random sample of size $n = 100$ is obtained from a population with $\mu = 300$ and $\sigma = 50$. Which of the following statements best describes the sampling distribution of $\bar{X}$ in terms of its shape, mean, and standard deviation?

(A) The sampling distribution of $\bar{X}$ is approximately normal with mean equal to 300 and standard deviation equal to 5.
(B) The sampling distribution of $\bar{X}$ is approximately normal with mean equal to 300 and standard deviation equal to 50.
(C) The sampling distribution of $\bar{X}$ is approximately normal with mean equal to 30 and standard deviation equal to 5.
(D) The sampling distribution of $\bar{X}$ is approximately normal with mean equal to 3 and standard deviation equal to 0.5.

316. A random sample of size $n = 36$ is obtained from a population with $\mu = 60$ and $\sigma = 15$. Which of the following statements best describes the sampling distribution of $\bar{X}$ in terms of its shape, mean, and standard deviation?

(A) The sampling distribution of $\bar{X}$ is approximately normal with mean equal to 10 and standard deviation equal to 15.
(B) The sampling distribution of $\bar{X}$ is approximately normal with mean equal to 60 and standard deviation equal to 2.5.
(C) The sampling distribution of $\bar{X}$ is approximately normal with mean equal to 60 and standard deviation equal to 15.
(D) The sampling distribution of $\bar{X}$ is positively skewed with mean equal to 10 and standard deviation equal to 2.5.

317. A random sample of size $n = 9$ is obtained from a normally distributed population with $\mu = 25$ and $\sigma = 6$. Which of the following statements best describes the sampling distribution of $\overline{X}$ in terms of its shape, mean, and standard deviation?

(A) The sampling distribution of $\overline{X}$ is a skewed distribution with mean equal to 25 and standard deviation equal to 6.
(B) The sampling distribution of $\overline{X}$ is a skewed distribution with mean equal to 25 and standard deviation equal to 2.
(C) The sampling distribution of $\overline{X}$ has a normal distribution with mean equal to 25 and standard deviation equal to 2.
(D) The sampling distribution of $\overline{X}$ has a normal distribution with mean equal to 25 and standard deviation equal to $\frac{2}{3}$.

318. A random sample of size $n = 100$ is obtained from a population with $\mu = 300$ and $\sigma = 50$. What is the probability that the sample mean will fall between 299 and 301?

(A) 0.0160
(B) 0.9545
(C) 0.0019
(D) 0.1585

319. A random sample of size $n = 36$ is obtained from a population with $\mu = 60$ and $\sigma = 15$. What is the probability that the sample mean will be at least 58?

(A) 0.7881
(B) 0.5530
(C) 0.4470
(D) 0.2119

320. A random sample of size $n = 9$ is obtained from a normally distributed population with $\mu = 25$ and $\sigma = 6$. What is the probability that the sample mean will exceed 26.7?

(A) 0.8023
(B) 0.1977
(C) 0.3885
(D) 0.6115

321. A random sample of size $n = 36$ is obtained from a population with $\mu =$ 30 and $\sigma = 12$. Find the probability that the sample mean will be no more than 26.08.

(A) 0.0250
(B) 0.9750
(C) 0.3720
(D) 0.6280

322. A random sample of size $n = 225$ is obtained from a population with $\mu = 120$ and $\sigma = 30$. Find the probability that the sample mean will lie between 122 and 124.

(A) 0.2645
(B) 0.0049
(C) 0.1359
(D) 0.0265

323. A random sample of size $n = 25$ is obtained from a normally distributed population with $\mu = 120$ and $\sigma = 15$. Find the probability that the sample mean will be no more than 126.

(A) 0.3446
(B) 0.9772
(C) 0.6554
(D) 0.0228

324. A random sample of size $n = 100$ is obtained from a population with $\mu = 1{,}200$ and $\sigma = 250$. Find the probability that the sample mean will lie between 1,200 and 1,250.

(A) 0.4772
(B) 0.9772
(C) 0.0793
(D) 0.1585

325. IQ scores for adults are normally distributed with mean $\mu = 100$ and standard deviation $\sigma = 15$. If a random sample of 16 adults is selected, find the probability that the average IQ score in the sample will be at least 92.5.

(A) 0.0228
(B) 0.9772
(C) 0.6915
(D) 0.3085

326. Suppose that scores on a national exam are normally distributed with mean $\mu = 500$ and standard deviation $\sigma = 100$. If a random sample of 200 test takers is selected, find the probability the average score of the sample will be between 500 and 520.

(A) 0.1000
(B) 0.0793
(C) 0.4977
(D) 0.5023

327. A local deli chain makes a signature sandwich. Suppose the measures of fat grams in the signature sandwiches are normally distributed with mean $\mu = 16$ grams and standard deviation $\sigma = 1.5$ grams. If a random sample of 4 signature sandwiches is selected, find the probability that the average amount of fat grams will be no more than 16.9.

(A) 0.1151
(B) 0.7257
(C) 0.2743
(D) 0.8849

For questions 328–332, obtain the confidence interval by formula. (Round your answers, as needed.)

328. A random sample of size $n = 50$ obtained from a population with unknown mean yielded $\overline{x} = 25$. Assume the standard deviation $\sigma = 8$. Construct a 95% confidence interval for μ.

(A) (22.8, 27.2)
(B) (23.1, 26.9)
(C) (11.8, 38.1)
(D) (9.32, 40.7)

329. A random sample of size $n = 9$ obtained from a normally distributed population with unknown mean μ yielded $\overline{x} = 26.7$. Assume the standard deviation $\sigma = 6$. Construct a 90% confidence interval for μ.

(A) (25.39, 28.01)
(B) (23.41, 29.99)
(C) (25.60, 27.80)
(D) (22.78, 30.62)

330. A random sample of size $n = 100$ obtained from a population with standard deviation $\sigma = 37$ and unknown mean μ yielded $\overline{x} = 301$. Construct a 90% confidence interval for μ.

(A) (291.5, 310.5)
(B) (300.4, 301.6)
(C) (294.9, 307.1)
(D) (293.8, 308.2)

331. A random sample of size $n = 36$ obtained from a population with unknown mean μ yielded $\overline{x} = 58$ and $s = 14$. Assume $\sigma = 15$. Construct a 99% confidence interval for μ.

(A) (52.0, 64.0)
(B) (53.9, 62.1)
(C) (57.3, 58.7)
(D) (51.6, 64.4)

332. A random sample of size $n = 225$ obtained from a population with unknown mean μ yielded $\overline{x} = 123$ and $s = 28$. Assume $\sigma = 25$. Construct a 90% confidence interval for μ.

(A) (119.7, 126.3)
(B) (120.3, 125.7)
(C) (118.7, 127.3)
(D) (119.9, 126.1)

For questions 333–337, obtain the specified confidence interval using a graphing calculator. (Round your answers, as needed.)

333. A researcher at a health science center wants to estimate the average amount of fat grams in the signature sandwich of a local deli. A random sample of 40 signature sandwiches yielded $\overline{x} = 17$ fat grams and $s = 2.9$ fat grams. Assume $\sigma = 2.5$ fat grams. Construct a 95% confidence interval for the average amount of fat grams in the signature sandwich of the local deli.

(A) (15.8, 18.2)
(B) (16.1, 17.9)
(C) (16.2, 17.8)
(D) (16.4, 17.7)

334. The fund-raising officer for a charity organization wants to estimate the average donation from contributors to the charity. A random sample of 100 donations yielded $\overline{x} = \$234.85$ and $s = \$98.35$. Assume $\sigma = \$95.23$. Construct a 90% confidence interval for the average donation from all contributors to the charity.

(A) $215.57 to $254.13
(B) $216.19 to $253.51
(C) $218.67 to $251.03
(D) $219.19 to $250.51

335. A manufacturer of car batteries needs to estimate the average life of its batteries. A random sample of 40 batteries had $\overline{x} = 39$ months and $s = 9$ months. Assume $\sigma = 8$ months. Calculate a 95% confidence interval for the average life of the manufacturer's batteries.

(A) 37.3 to 40.8 months
(B) 36.4 to 41.6 months
(C) 36.5 to 41.5 months
(D) 36.2 to 41.8 months

336. In a study of 500 randomly selected adults in the United States, the average calorie consumption of participants was 2,400 calories per day. Assume a population standard deviation of 930 calories per day. Calculate a 95% confidence interval for the true mean caloric intake per day of all adults in the United States.

(A) 2,292.9 to 2,507.1 calories
(B) 2,318.5 to 2,481.5 calories
(C) 2,331.6 to 2,468.4 calories
(D) 2,396.4 to 2,403.6 calories

337. A real estate broker needs to know the average house price in a metropolitan area. A random sample of 50 house prices in the metropolitan area yielded $\overline{x}$ = \$120,000. Assume σ = \$47,000. Give an 80% confidence interval for the average house price in the metropolitan area.

(A) \$106,972 to \$133,028
(B) \$106,643 to \$133,357
(C) \$111,482 to \$128,518
(D) \$111,365 to \$128,635

For questions 338–344, obtain the specified confidence interval by formula. Use the *t*-distribution table in Appendix C to find the *t* values. (Round your answers, as needed.)

338. Consider a random sample of 16 measurements obtained from a normally distributed population with $\overline{x} = 450$ and $s = 60$. Construct a 90% confidence interval.

(A) (423.7, 476.3)
(B) (429.9, 470.1)
(C) (430.8, 469.2)
(D) (447.3, 452.6)

339. Consider a random sample of 9 measurements obtained from a normally distributed population with $\overline{x} = 450$ and $s = 60$. Construct a 90% confidence interval.

(A) (417.1, 482.9)
(B) (412.8, 487.2)
(C) (403.9, 496.1)
(D) (448.1, 451.9)

340. Consider a random sample of 25 measurements obtained from a normally distributed population with $\overline{x} = 450$ and $s = 60$. Construct a 90% confidence interval.

(A) (425.2, 474.8)
(B) (430.2, 469.7)
(C) (429.5, 470.5)
(D) (444.6, 455.4)

341. Consider a random sample of 4 measurements obtained from a normally distributed population with $\overline{x} = 450$ and $s = 60$. Construct a 95% confidence interval.

(A) (354.5, 545.5)
(B) (391.2, 508.8)
(C) (379.4, 520.6)
(D) (449.0, 451.0)

342. Consider a random sample of 25 measurements obtained from a normally distributed population with $\overline{x} = 450$ and $s = 60$. Construct a 95% confidence interval.

(A) (429.5, 470.5)
(B) (425.2, 474.8)
(C) (426.5, 473.5)
(D) (443.5, 456.5)

343. Consider a random sample of 16 measurements obtained from a normally distributed population with $\overline{x} = 450$ and $s = 60$. Construct a 99% confidence interval.

(A) (423.7, 476.3)
(B) (420.1, 479.9)
(C) (405.8, 494.2)
(D) (411.4, 488.6)

344. Consider a random sample of 25 measurements obtained from a normally distributed population with $\overline{x} = 450$ and $s = 60$. Construct a 99% confidence interval.

(A) (449.0, 511.0)
(B) (419.1, 480.9)
(C) (425.2, 474.8)
(D) (416.4, 483.6)

For questions 345–349, obtain the specified confidence interval using a graphing calculator. (Round your answers, as needed.)

345. Suppose that scores on a standardized exam are normally distributed. To estimate the average score on the exam for all test takers, a researcher obtained a random sample of 25 scores. The sample yielded $\overline{x} = 480$ and $s = 75$. Give a 95% confidence interval for the average score on the exam for all test takers.

(A) (454.3, 505.7)
(B) (449.0, 511.0)
(C) (420.1, 479.9)
(D) (474.3, 485.7)

346. A local deli chain advertises a "heart-healthy" sandwich. Suppose the number of fat grams in a heart-healthy sandwich is a normal random variable. To estimate the average amount of fat grams in the deli's heart-healthy sandwiches, a nutritionist selected a random sample of 4 heart-healthy sandwiches. The sample yielded $\overline{x} = 11$ grams and $s = 1.8$ grams. Construct a 90% confidence interval for the average amount of fat grams in a heart-healthy sandwich.

(A) (8.1, 13.9)
(B) (8.9, 13.1)
(C) (5.7, 16.3)
(D) (9.5, 12.5)

347. A plumbing company has found that the length of time, in minutes, for installing a bathtub is normally distributed. The owner of the company hired a statistician to estimate the average length of time the company takes to install a bathtub. The statistician randomly selected a sample of 10 bathtub installations. The sample yielded $\overline{x} = 150$ minutes and $s =$ 30 minutes. Construct a 95% confidence interval for the average length of time the company takes to install a bathtub.

(A) (131.4, 168.6)
(B) (134.4, 165.6)
(C) (128.5, 171.5)
(D) (132.6, 167.4)

348. The weights of newborn baby girls born at a local hospital are normally distributed. A hospital administrator conducted a study to estimate the mean weight of all newborn baby girls born at the hospital. A random sample of size $n = 20$ had a mean weight of 95 ounces and a standard deviation of 8 ounces. Calculate a 99% confidence interval for the mean weight of all newborn baby girls born at the hospital.

(A) (89.9, 100.1)
(B) (91.3, 98.7)
(C) (91.9, 98.1)
(D) (90.4, 99.6)

349. The lifetime for a certain brand of incandescent light bulbs is normally distributed. A random sample of 5 of these light bulbs had a mean lifetime of 520 hours and standard deviation of 50 hours. Give a 95% confidence interval for the mean lifetime of all incandescent light bulbs of this brand.

(A) (476.2, 563.8)
(B) (417.1, 622.9)
(C) (472.3, 567.7)
(D) (457.9, 582.1)

For questions 350–352, obtain the specified confidence interval by formula. (Round your answers, as needed.)

350. $n = 200$, $\hat{p} = 0.2$, 90% confidence interval

(A) (0.1446, 0.2554)
(B) (0.0651, 1349)
(C) (0.1535, 0.2465)
(D) (0.1271, 2729)

351. $n = 100$, $\hat{p} = 0.45$, 99% confidence interval

(A) (0.3218, 0.5782)
(B) (0.3682, 0.5318)
(C) (0.3525, 5475)
(D) (0.3862, 0.5138)

352. $n = 1{,}500$, $x = 45$, 95% confidence interval

(A) (0.0228, 0.0372)
(B) (0.0214, 0.0386)
(C) (0.0187, 0.0413)
(D) (0.0198, 0.0402)

353. For which of the following conditions is the stated sample size n large enough for constructing a confidence interval for p?

(A) $n = 20, \hat{p} = 0.2$
(B) $n = 100, \hat{p} = 0.03$
(C) $n = 15, \hat{p} = 0.6$
(D) $n = 150, \hat{p} = 0.99$

For questions 354 and 355, obtain the margin of error for the specified confidence interval by formula. (Round your answers, as needed.)

354. $n = 15, \hat{p} = 0.6$, 90% confidence interval

(A) 0.2479
(B) 0.3258
(C) 0.1621
(D) 0.2081

355. $n = 1{,}000, \hat{p} = 0.82$, 95% confidence interval

(A) 0.0238
(B) 0.0313
(C) 0.0200
(D) 0.0154

For questions 356–360, obtain the specified confidence interval using a graphing calculator.

356. A random check of 500 batteries manufactured by a company found 15 defective batteries. Construct a 95% confidence interval for the proportion of all the company's batteries that are defective.

(A) (0.01505, 0.04495)
(B) (0.01902, 0.04098)
(C) (0.017545, 0.04255)
(D) (0.01035, 0.04965)

357. Assume a study revealed that from a sample of 80 working women in a metropolitan area, 12 were over the age of 60. Construct a 99% confidence interval for the true proportion of working women over the age of 60 in the metropolitan area.

(A) (0.07175, 0.22825)
(B) (0.04717, 0.25283)
(C) (0.08433, 0.21567)
(D) (0.09884, 0.20116)

358. A radio station's advertising manager needs to know the proportion of radio listeners in the area who listen to the manager's radio station. Suppose that in a random sample of 100 radio listeners in the area, 34 listen to the manager's station. Construct a 90% confidence interval for the proportion of radio listeners in the area who listen to the manager's radio station.

(A) (0.24715, 0.43285)
(B) (0.21798, 0.46202)
(C) (0.26208, 0.41792)
(D) (0.27181, 0.40819)

359. An interior designer needs to know the proportion of assisted-living residents who prefer blue for the color of their bedroom walls. In a random survey of 300 assisted-living residents, 80 percent said they prefer blue. Construct a 95% confidence interval for the proportion of all assisted-living residents who prefer blue on bedroom walls.

(A) (0.76201, 0.83799)
(B) (0.74071, 0.85949)
(C) (0.77040, 0.82960)
(D) (0.75474, 0.84526)

360. A market researcher wants to know the percent of households in a rural area that have no vehicle. In a random sample of 120 households, 6 households had no vehicle. Construct a 90% confidence interval for the percent of households in the rural area that have no vehicle.

(A) (0.01727, 0.08273)
(B) (–0.00120, 0.10125)
(C) (0.01101, 0.08899)
(D) (0.02711, 0.07289)

361. Find the sample size necessary to estimate the population mean μ to within the margin of error $E = 2$ with 95% confidence, given that prior data suggest $\sigma = 10$.

(A) 96
(B) 95
(C) 97
(D) 99

362. Find the sample size necessary to estimate the population mean μ to within the margin of error $E = 2$ with 95% confidence, given that prior data suggest $\sigma = 100$.

(A) 9,604
(B) 9,603
(C) 960
(D) 961

363. Find the sample size necessary to estimate the population mean μ to within the margin of error $E = 0.2$ with 95% confidence, given that prior data suggest $\sigma = 100$.

(A) 960,401
(B) 960,400
(C) 960,000
(D) 960,500

364. Find the sample size necessary to estimate the population mean μ to within the margin of error $E = 1.5$ with 95% confidence, given that prior data suggest $\sigma = 5$.

(A) 420
(B) 42
(C) 43
(D) 430

365. Find the sample size necessary to estimate the population mean μ to within the margin of error $E = 0.15$ with 95% confidence, given that prior data suggest that $\sigma = 5$.

(A) 4,267
(B) 42,670
(C) 4,269
(D) 4,268

366. Find the sample size necessary to estimate the population proportion p to within the margin of error $E = 0.02$ with 95% confidence, given that prior data suggest $p = 0.1$.

(A) 865
(B) 864
(C) 870
(D) 860

367. Find the sample size necessary to estimate the population proportion p to within the margin of error $E = 0.02$ with 95% confidence, given that prior data suggest $p = 0.2$.

(A) 1,536
(B) 1,537
(C) 1,538
(D) 1,539

368. Find the sample size necessary to estimate the population proportion p to within the margin of error $E = 0.02$ with 95% confidence, given that prior data suggest $p = 0.5$.

(A) 2,400
(B) 2,500
(C) 2,402
(D) 2,401

369. Find the sample size necessary to estimate the population proportion p to within the margin of error $E = 0.02$ with 95% confidence, given that prior data suggest $p = 0.7$.

(A) 2,017
(B) 2,016
(C) 2,020
(D) 2,010

370. Find the sample size necessary to estimate the population proportion p to within the margin of error $E = 0.02$ with 95% confidence, given that prior data suggest $p = 0.9$.

(A) 864
(B) 865
(C) 866
(D) 867

371. A health professional wants to estimate the birth weights of infants. A previous study indicated that the standard deviation of infant birth weights is 7.6 ounces. What size sample is necessary if the health professional wants to estimate the true mean birth weights of infants to within 1.5 ounces with 99% confidence?

(A) 170
(B) 171
(C) 180
(D) 181

372. A pollster wants to conduct a survey to estimate the proportion of US citizens who favor a certain presidential candidate within a margin of error $E = 0.04$ ("4 percentage points") with 95% confidence. Assuming no prior information is available, find the minimum required sample size for the pollster's survey.

(A) 60,025
(B) 60,026
(C) 600
(D) 601

373. A manufacturer of car batteries needs to estimate the average life of its batteries. Prior data indicate that the standard deviation of the population of batteries is approximately 8 months. What size sample is necessary if the manufacturer wants to estimate the true average life of its car batteries within 2 months with 90% confidence?

(A) 44
(B) 45
(C) 43
(D) 46

374. The fund-raising officer for a charity organization wants to estimate the average donation from contributors to the charity. Prior data indicate that the standard deviation of charitable contributions to the organization is approximately $95. What size sample is necessary if the fund-raising officer wants to estimate the true average donation from all contributors to the charity within $20 with 95% confidence?

(A) 86
(B) 87
(C) 88
(D) 89

375. A quality-control inspector for a company that makes cell phones needs to estimate the proportion of defective cell phones produced by the company. The quality-control inspector estimates that the proportion defective is about 0.1, corresponding to 10% defective. How many of the company's cell phones should be sampled and checked in order to estimate the proportion of defective cell phones to within 0.01 with 90% confidence?

(A) 2,400
(B) 2,435
(C) 2,436
(D) 2,500

CHAPTER 11

One-Sample Hypothesis Tests of Means and Proportions

376. Which one of the following statements is always true?

(A) A one-sample test of hypothesis has two competing hypotheses about a population statistic.

(B) In a test of hypothesis, only the alternative hypothesis is actually tested.

(C) In formulating hypotheses, no information from the sample is used in the hypotheses statements.

(D) A helpful guideline for formulating hypotheses is that the null hypothesis always contains an inequality.

377. Which one of the following statements is always true?

(A) If H_0 contains $\neq$, the hypothesis test is two-tailed.

(B) If H_a contains $>$, the hypothesis test is left-tailed.

(C) If H_a contains $\neq$, the hypothesis test is two-tailed.

(D) If H_0 contains $\leq$, the hypothesis test is left-tailed.

378. Which one of the following statements is always true?

(A) If a test of hypothesis is significant at the 5% significance level, then the test would also be significant at the 1% significance level.

(B) If a test of hypothesis is significant at the 1% significance level, then the test would also be significant at the 5% significance level.

(C) If a test of hypothesis is significant at the 10% significance level, then the test would also be significant at the 1% significance level.

(D) If a test of hypothesis is significant at the 10% significance level, then the test would also be significant at the 5% significance level.

379. Which one of the following statements is always true?

(A) The greater the *p*-value of a hypothesis test, the stronger is the belief in the null hypothesis.
(B) If the *p*-value of a hypothesis test is less than α, its significance level, the null hypothesis should be rejected.
(C) The *p*-value of a hypothesis test is the probability that the null hypothesis is true.
(D) The significance level is the probability of rejecting a false null hypothesis.

380. A manufacturer of car batteries claims the average life of its batteries is at least 45 months. A random sample of 80 car batteries from the manufacturer yielded $\bar{x} = 39$ months and $s = 8$ months. A consumer group wants to determine whether the sample data dispute the manufacturer's claim. Assuming the population is normally distributed σ, which inferential statistical test is most appropriate for this situation?

(A) one-sample *t*-test
(B) one-sample *z*-test
(C) one-sample proportion *z*-test
(D) none of the above

381. The fund-raising officer for a charity organization claims the average donation from contributors to the charity is \$250.00. To test the claim, a random sample of 100 donations is obtained, yielding $\bar{x} = \$234.85$ and $s = \$98.35$. Assume $\sigma = \$95.23$. Which inferential statistical test is most appropriate for this situation?

(A) one-sample *t*-test
(B) one-sample *z*-test
(C) one-sample proportion *z*-test
(D) none of the above

382. A local deli chain advertises that its "heart-healthy" sandwich contains a mere 10 fat grams. To test the deli chain's claim, a nutritionist selected a random sample of 4 heart-healthy sandwiches. The sample yielded $\bar{x} = 11$ grams and $s = 1.8$ grams. Assuming σ is unknown, which inferential statistical test is most appropriate for this situation?

(A) one-sample *t*-test
(B) one-sample *z*-test
(C) one-sample proportion *z*-test
(D) none of the above

383. In a sample for a government study of 80 working women in a metropolitan area, 30 were over the age of 60. Government officials want to know whether the sample data dispute an earlier claim that the true proportion of working women over the age of 60 in the metropolitan area is no more than 0.18. Which inferential statistical test is most appropriate for this situation?

(A) one-sample t-test
(B) one-sample z-test
(C) one-sample proportion z-test
(D) none of the above

384. A real estate broker's newsletter says the average price of a house in a metropolitan area is \$300,000. A random sample of 50 house prices in the metropolitan area yielded $\bar{x}$ = \$320,000. Assume σ = \$47,000. You want to know whether the sample data cast doubt on the average given in the broker's newsletter. Which inferential statistical test is most appropriate?

(A) one-sample t-test
(B) one-sample z-test
(C) one-sample proportion z-test
(D) none of the above

385. A professor of interior design conjectures that the proportion of assisted-living residents who would prefer blue for their bedroom walls is greater than 0.75. In a random sample of 300 assisted-living residents, 240 said they preferred blue on bedroom walls. The professor wants to determine whether the sample data support the conjecture. Which inferential statistical test is most appropriate?

(A) one-sample t-test
(B) one-sample z-test
(C) one-sample proportion z-test
(D) none of the above

386. The weights of newborn baby girls born at a county hospital are normally distributed. A hospital administrator conducts a study to determine whether the mean weight of all newborn baby girls born at the hospital is less than 100 ounces. A random sample of size $n = 20$ had a mean weight of 95 ounces and a standard deviation of 7 ounces. Assume σ is unknown. To determine whether, at the 1% significance level, the data indicate that the mean weight of all newborn baby girls born at the county hospital is less than 100 ounces, which inferential statistical test is most appropriate?

(A) one-sample t-test
(B) one-sample z-test
(C) one-sample proportion z-test
(D) none of the above

387. A random check of 500 lightning cables manufactured by a company found 15 were defective. Quality-control guidelines established by the manufacturer require that the proportion of its lightning cables that are defective must be less than 0.05. The manufacturer wants to determine whether it is meeting its quality control guidelines. Which inferential statistical test is most appropriate for this situation?

(A) one-sample *t*-test
(B) one-sample *z*-test
(C) one-sample proportion *z*-test
(D) none of the above

388. The average caloric intake of 500 randomly selected adults in the United States was 2,400 calories per day. The researchers want to test a prior claim that the true mean caloric intake of all adults in the United States is no more than 2,300 calories per day. Assume a population standard deviation of 930 calories. Which inferential statistical test is most appropriate for this situation?

(A) one-sample *t*-test
(B) one-sample *z*-test
(C) one-sample proportion *z*-test
(D) none of the above

389. The advertising manager for a radio station wants to test the hypothesis that the percent of radio listeners in the area who listen to the manager's radio station is at least 40%. In a random sample of 100 radio listeners in the area, 34 said they listen to the manager's station. Which inferential statistical test is most appropriate for testing the manager's hypothesis?

(A) one-sample *t*-test
(B) one-sample *z*-test
(C) one-sample proportion *z*-test
(D) none of the above

390. In a random sample of 100 remote controls manufactured by a company, 2 were found to be defective. The manufacturer's quality-control guidelines require that the percent of all its remote controls that are defective must be less than 3%. The manufacturer wants to determine whether it is meeting its quality control guidelines. Which inferential statistical test is most appropriate for this situation?

(A) one-sample *t*-test
(B) one-sample *z*-test
(C) one-sample proportion *z*-test
(D) none of the above

391. The fund-raising officer for a charity organization claims the average donation from contributors to the charity is \$250.00. To test the claim, a random sample of 100 donations is obtained, yielding $\overline{x}$ = \$234.85 and s = \$98.35. Assume σ = \$95.23. State the null and alternative hypotheses for the appropriate hypothesis test.

(A) $H_0 : \mu = 250.00; H_a : \mu \neq 250.00$
(B) $H_0 : \mu \neq 250.00; H_a : \mu = 250.00$
(C) $H_0 : \mu = \$234.85; H_a : \mu \neq \$234.85$
(D) $H_0 : \mu \neq \$234.85; H_a : \mu = \$234.85$

392. A manufacturer of car batteries claims the average life of its batteries is at least 45 months. A random sample of 80 car batteries from the manufacturer yielded $\overline{x} = 39$ months and $s = 8$ months. A consumer group wants to determine whether the sample data dispute the manufacturer's claim. Assuming the population is normally distributed and σ is unknown, state the null and alternative hypotheses for the appropriate hypothesis test.

(A) $H_0 : \mu \geq 39; H_a : \mu < 39$
(B) $H_0 : \mu \geq 45; H_a : \mu < 45$
(C) $H_0 : \mu < 39; H_a : \mu \geq 39$
(D) $H_0 : \mu < 45; H_a : \mu \geq 45$

393. The average caloric intake of 500 randomly selected adults in the United States was 2,400 calories per day. The researchers want to test an earlier claim that the true mean caloric intake of all adults in the United States is no more than 2,300 calories per day. Assume a population standard deviation of 930 calories. State the null and alternative hypotheses for the appropriate hypothesis test.

(A) $H_0 : \mu \leq 2{,}400; H_a : \mu > 2{,}400$
(B) $H_0 : \mu > 2{,}400; H_a : \mu \leq 2{,}400$
(C) $H_0 : \mu \leq 2{,}300; H_a : \mu > 2{,}300$
(D) $H_0 : \mu > 2{,}300; H_a : \mu \leq 2{,}300$

394. A local deli chain advertises that its "heart-healthy" sandwich has a mere 10 fat grams. Suppose that the amount of fat grams in a heart-healthy sandwich is a normal random variable and σ is unknown. To test the deli chain's claim, a nutritionist selected a random sample of 4 heart-healthy sandwiches. The sample yielded $\overline{x} = 11$ grams and $s = 1.8$ grams. State the null and alternative hypotheses for the appropriate hypothesis test.

(A) $H_0 : \mu = 11; H_a : \mu < 11$
(B) $H_0 : \mu = 11; H_a : \mu \neq 11$
(C) $H_0 : \mu = 10; H_a : \mu < 10$
(D) $H_0 : \mu = 10; H_a : \mu \neq 10$

395. A real estate broker's newsletter says the average price of a house in a metropolitan area is \$300,000. A random sample of 50 house prices in the metropolitan area yielded $\bar{x} = \$320{,}000$. Assuming $\sigma = \$47{,}000$, you want to know whether the sample data cast doubt on the listing in the broker's newsletter. State the null and alternative hypotheses for the appropriate hypothesis test.

(A) $H_0 : \mu = 300{,}000; H_a : \mu \neq 300{,}000$
(B) $H_0 : \mu = 320{,}000; H_a : \mu \neq 300{,}000$
(C) $H_0 : \mu = 300{,}000; H_a : \mu > 300{,}000$
(D) $H_0 : \mu = 320{,}000; H_a : \mu > 300{,}000$

396. The weights of newborn baby girls born at a county hospital are normally distributed. A hospital administrator conducts a study to determine whether the mean weight of all newborn baby girls born at the hospital is less than 100 ounces. A random sample of size $n = 20$ had a mean weight of 95 ounces and a standard deviation of 7 ounces. Assume σ is unknown. The administrator's research question is whether, at the 1% significance level, these data indicate that the mean weight of all newborn baby girls born at the county hospital is less than 100 ounces. State the null and alternative hypotheses for the appropriate hypothesis test.

(A) $H_0 : \mu \leq 100; H_a : \mu < 100$
(B) $H_0 : \mu \geq 100; H_a : \mu < 100$
(C) $H_0 : \mu \geq 95; H_a : \mu < 95$
(D) $H_0 : \mu \leq 95; H_a : \mu > 95$

397. A random check of 500 lightning cables manufactured by a company found 15 were defective. The manufacturer's quality-control guidelines require that the proportion of its lightning cables that are defective must be less than 0.05. The manufacturer wants to determine whether it is meeting its quality control guidelines. State the null and alternative hypotheses for the appropriate hypothesis test.

(A) $H_0 : p \geq 15; H_a : p < 15$
(B) $H_0 : p = 15; H_a : p \neq 15$
(C) $H_0 : p \geq 0.05; H_a : p < 0.05$
(D) $H_0 : p = 0.05; H_a : p \neq 0.05$

398. In a sample for a government study of 80 working women in a metropolitan area, 30 were over the age of 60. Government officials want to know whether the sample data dispute an earlier claim that the true proportion of working women over the age of 60 in the metropolitan area is no more than 0.18. State the null and alternative hypotheses for the appropriate hypothesis test.

(A) $H_0 : p = 80; H_a : p > 80$
(B) $H_0 : p \leq 80; H_a : p > 80$
(C) $H_0 : p \leq 0.18; H_a : p > 0.18$
(D) $H_0 : p > 0.18; H_a : p \leq 0.18$

399. The advertising manager for a radio station wants to test the hypothesis that the percent of radio listeners in the area who listen to the manager's radio station is at least 40%. In a random sample of 100 radio listeners in the area, 34 said they listen to the manager's station. State the null and alternative hypotheses for the appropriate hypothesis test.

(A) $H_0: p \geq 0.40$; $H_a: p < 0.40$
(B) $H_0: p = 0.40$; $H_a: p = 0.60$
(C) $H_0: p \geq 0.34$; $H_a: p < 0.34$
(D) $H_0: p = 0.34$; $H_a: p = 0.66$

400. A professor of interior design conjectures that the proportion of assisted-living residents who prefer blue for their bedroom walls is greater than 0.75. In a random sample of 300 assisted-living residents, 240 said they prefer blue on bedroom walls. The professor wants to determine whether the sample data support the conjecture. State the null and alternative hypotheses for the appropriate hypothesis test.

(A) $H_0: p > 0.75$; $H_a: p < 0.75$
(B) $H_0: p \geq 0.75$; $H_a: p < 0.75$
(C) $H_0: p \leq 0.75$; $H_a: p > 0.75$
(D) $H_0: p < 0.75$; $H_a: p > 0.75$

401. Suppose that, unknown to a researcher, the reality of the following hypotheses is as shown. The researcher decides to accept H_0. Which of the following statements is true about the decision?

$H_0: \mu \geq 15$	False
$H_a: \mu < 15$	True

(A) The researcher made the correct decision to accept H_0.
(B) The researcher made the correct decision to reject H_a.
(C) The researcher made a type I error.
(D) The researcher made a type II error.

402. Suppose that, unknown to a researcher, the reality of the following hypotheses is as shown. The researcher decides to reject H_0 and accept H_a. Which of the following statements is true about the decision?

$H_0: \mu \leq 500$	False
$H_a: \mu > 500$	True

(A) The researcher made the correct decision to accept H_a.
(B) The researcher made the wrong decision to reject H_0.
(C) The researcher made a type I error.
(D) The researcher made a type II error.

403. Suppose that, unknown to a researcher, the reality of the following hypotheses is as shown. The researcher decides to reject H_0 and accept H_a. Which of the following statements is true about the decision?

$H_0 : p = 0.60$	True
$H_a : p \neq 0.60$	False

(A) The researcher made the correct decision to accept H_a.
(B) The researcher made the correct decision to reject H_0.
(C) The researcher made a type I error.
(D) The researcher made a type II error.

404. Which one of the following statements is always true?
(A) The probability of making a type II error is less than the probability of making a type I error.
(B) The probability of making a type I error and the probability of making a type II error are inversely related for a fixed sample size.
(C) For a given sample size, as the probability of making a type I error increases, the probability of making a type II error increases as well.
(D) Researchers customarily set the risk of a type I error at 0.95.

405. The hypotheses for a hypothesis test concerning μ from a normally distributed population are as shown:

$H_0 : \mu \geq 15$

$H_a : \mu < 15$

Suppose $n = 20$, and σ is known. At the 5% significance level, H_0 will be rejected if and only if the test statistic is
(A) less than −1.645
(B) less than −1.645 or greater than 1.645
(C) less than −1.729
(D) less than −1.729 or greater than 1.729

406. The hypotheses for a hypothesis test concerning p are as follows:

$H_0 : p = .8$

$H_a : p \neq .8$

Suppose $n = 400$. At the 5% significance level, H_0 will be rejected if and only if the test statistic is
(A) less than −1.645
(B) less than −1.645 or greater than 1.645
(C) less than −1.96
(D) less than −1.96 or greater than 1.96

407. The hypotheses for a hypothesis test concerning μ are as follows:

$H_0: \mu \leq 12$
$H_a: \mu > 12$

Suppose $n = 35$, and σ is unknown. At the 5% significance level, H_0 will be rejected if and only if the test statistic is

(A) greater than 1.645
(B) less than −1.645 or greater than 1.645
(C) greater than 1.691
(D) less than −1.691 or greater than 1.691

For questions 408–412, compute the test statistic (*T.S.*) for the appropriate hypothesis test by formula. (Round your answers, as needed.)

408. To test a claim that the mean of a certain population is 22 at the 5% significance level, a random sample of size $n = 50$ was obtained from the population. The sample yielded $\overline{x} = 25$. Assume the standard deviation $\sigma = 8$. Compute *T.S.*

(A) 2.652
(B) 0.375
(C) −0.375
(D) −2.652

409. A random sample of size $n = 40$ obtained from a normally distributed population with unknown σ yielded $\overline{x} = 26.7$ and standard deviation $s = 6$. The researcher wants to know whether the data cast doubt on a conjecture that the population mean is at least 28. Compute *T.S.*

(A) −0.217
(B) −1.370
(C) 1.370
(D) 0.217

410. To test a claim that the mean of a certain population is greater than 300, researchers obtain a random sample of size $n = 100$ from the population, yielding $\overline{x} = 301$ and $s = 37$. The researchers want to determine at the 5% significance level whether these data support the claim. Compute *T.S.*

(A) −0.270
(B) −0.027
(C) 0.270
(D) 0.027

411. A random sample of size $n = 36$ obtained from a population with standard deviation $\sigma = 15$ yielded $\overline{x} = 58$ and $s = 14$. Before obtaining the sample, the researcher had conjectured that the mean was no more than 50. A test of hypothesis is performed at the 1% significance level to test the conjecture. Compute *T.S.*

(A) −3.429
(B) 3.429
(C) −3.200
(D) 3.200

412. To investigate a claim that the mean of a population exceeds 120, a random sample of size $n = 225$ was obtained from the population, yielding $\overline{x} = 123$. Assume $\sigma = 25$. The research team wants to know whether the sample data support the claim at the 5% significance level. Compute *T.S.*

(A) 1.800
(B) 0.120
(C) −1.800
(D) −0.120

For questions 413–417, compute the test statistic (*T.S.*) by using a graphing calculator. (Round your answers, as needed.)

413. The fund-raising officer for a charity organization claims the average donation from contributors to the charity is \$250.00. To test the claim, a random sample of 100 donations is obtained, yielding $\overline{x} = \$234.85$ and $s = \$98.35$. Assume $\sigma = \$95.23$. Compute *T.S.*

(A) −1.540
(B) −1.591
(C) 1.591
(D) 1.540

414. A manufacturer of car batteries claims the average life of its batteries is at least 45 months. A random sample of 80 car batteries from the manufacturer yielded $\overline{x} = 39$ months, and $s = 8$ months. A consumer group wants to determine whether the sample data dispute the manufacturer's claim. Assuming the population is normally distributed and σ is unknown, compute *T.S.*

(A) 6.708
(B) 9.910
(C) −6.708
(D) −9.910

415. The average caloric intake of 500 randomly selected adults in the United States was 2,400 calories per day. The researchers want to test a prior claim that the true mean caloric intake of all adults in the United States is no more than 2,300 calories per day. Assume a population standard deviation of 930 calories. Compute *T.S.*

(A) −0.008
(B) −2.404
(C) 0.008
(D) 2.404

416. A real estate broker's newsletter says the average price of a house in a metropolitan area is $300,000. A random sample of 50 house prices in the metropolitan area yielded $\overline{x}$ = $320,000. Assume σ = $47,000. To test whether, at the 1% significance level, these data cast doubt on the average in the broker's newsletter, compute *T.S.*

(A) 3.009
(B) −3.009
(C) 0.003
(D) −0.003

417. The weights of newborn baby girls born at a county hospital are normally distributed. A hospital administrator conducts a study to determine whether the mean weight of all newborn baby girls born at the hospital is less than 100 ounces. A random sample of size $n = 20$ had a mean weight of 95 ounces and a standard deviation of 7 ounces. Assume σ is unknown. To determine whether, at the 1% significance level, these data indicate that the mean weight of all newborn baby girls born at the county hospital is less than 100 ounces, compute *T.S.*

(A) 0.002
(B) −3.194
(C) −0.002
(D) 3.194

418. At the 5% significance level, for which one of the following *p*-values will H_0 be rejected?

(A) *p*-value = 0.3498
(B) *p*-value = 0.0499
(C) *p*-value = 0.0501
(D) *p*-value = 0.2000

For questions 419–420, determine the p-value for the appropriate hypothesis test by using a graphing calculator. (Round your answers, as needed.)

419. A claim is made that the mean of a certain population is 22. To test the claim at the 5% significance level, a random sample of size $n = 50$ was obtained from the population, yielding $\bar{x} = 25$. Assume the standard deviation $\sigma = 8$. Find the p-value for the test of hypothesis.

(A) 0.0040
(B) 0.0250
(C) 0.0080
(D) 0.9960

420. A random sample of size $n = 40$ obtained from a normally distributed population with unknown σ yielded $\bar{x} = 26.7$ and standard deviation $s = 6$. The researcher wants to know, at the 1% significance level, whether the data cast doubt on a conjecture that the population mean is at least 28. Find the p-value for the test of hypothesis.

(A) 0.9108
(B) 0.1784
(C) 0.0100
(D) 0.0892

421. A claim is made that the mean of a certain population is 22. To test the claim at the 5% significance level, a random sample of size $n = 50$ was obtained from the population, yielding $\bar{x} = 25$. Assume the standard deviation $\sigma = 8$. State the conclusion for the test of hypothesis.

(A) At the 5% significance level, the data provide sufficient evidence to indicate that the mean of the population is not 22.
(B) At the 5% significance level, the data provide sufficient evidence to indicate that the mean of the population is 22.
(C) At the 5% significance level, the data do not provide sufficient evidence to indicate that the mean of the population is not 22.
(D) At the 5% significance level, the data do not provide sufficient evidence to indicate that the mean of the population is 25.

422. A random sample of size $n = 40$ obtained from a normally distributed population with unknown σ yielded $\bar{x} = 26.7$ and standard deviation $s = 6$. The researcher wants to know, at the 1% significance level, whether the data cast doubt on a conjecture that the population mean is at least 28. State the conclusion for the test of hypothesis.

(A) At the 1% significance level, the hypothesis that the mean of the population is less than or equal to 28 is supported by the data.
(B) At the 1% significance level, the hypothesis that the mean of the population is greater than 28 is not supported by the data.
(C) At the 1% significance level, the hypothesis that the mean of the population is less than 28 is not supported by the data.
(D) At the 1% significance level, the hypothesis that the mean of the population is less than 28 is supported by the data.

For questions 423–432, state the correct decision rule for the test of hypothesis in terms of a *z*- or *t*-test statistic and appropriate rejection region at the indicated significance level.

423. A claim is made that the mean of a certain normally distributed population is 60. To test the claim, a research team obtained a random sample of size $n = 16$ from the population. The sample yielded $\bar{x} = 50$ and a standard deviation $s = 4$. Assume σ is unknown. State the decision rule at the 5% significance level.

(A) Reject H_0 if either $T.S. < -1.96$ or $T.S. > 1.96$.
(B) Reject H_0 if either $T.S. < -2.131$ or $T.S. > 2.131$.
(C) Reject H_0 if either $T.S. < -1.753$ or $T.S. > 1.753$.
(D) It cannot be determined from the information given.

424. A random sample of size $n = 25$ obtained from a normally distributed population with unknown mean μ and known $\sigma = 5$ yielded $\bar{x} = 26.7$ and standard deviation $s = 6.5$. Before obtaining the sample, the researcher conjectured that the population mean was at least 28. State the decision rule at the 5% significance level.

(A) Reject H_0 if $T.S. < -1.645$.
(B) Reject H_0 if $T.S. < -1.96$.
(C) Reject H_0 if $T.S. < -1.711$.
(D) It cannot be determined from the information given.

425. A claim is made that the mean of a certain normally distributed population is greater than 300. A random sample of size $n = 15$ obtained from the population yielded $\bar{x} = 301$ and $s = 37$. Assume σ is unknown. For determining whether the sample data support the claim, state the decision rule at the 10% significance level.

(A) Reject H_0 if $T.S. > 1.345$.
(B) Reject H_0 if $T.S. > 1.761$.
(C) Reject H_0 if $T.S. > 1.282$.
(D) It cannot be determined from the information given.

426. A random sample of size $n = 36$ obtained from a population with standard deviation $\sigma = 15$ yielded $\bar{x} = 58$ and $s = 14$. Before obtaining the sample, the researcher conjectured that the population mean was no more than 50. To test the conjecture, the researcher performed a test of hypothesis at the 1% significance level. State the decision rule.

(A) Reject H_0 if $T.S. > 2.576$.
(B) Reject H_0 if $T.S. < -2.576$ or $T.S. > 2.576$.
(C) Reject H_0 if $T.S. > 2.326$.
(D) It cannot be determined from the information given.

427. A claim is made that the mean of a normally distributed population exceeds 120. A random sample of size $n = 20$ obtained from the population yielded $\bar{x} = 123$ and $s = 25$. Assume σ is unknown. For a hypothesis test of whether the sample data support the claim, state the decision rule at the 5% significance level.

(A) Reject H_0 if $T.S. > 1.729$.
(B) Reject H_0 if $T.S. > 2.093$.
(C) Reject H_0 if $T.S. > 1.325$.
(D) It cannot be determined from the information given.

428. Suppose that scores on a standardized exam are normally distributed with $\sigma = 100$. The test company that administers the exam claims that the average score on the exam for all test takers is 500. To test the claim, a researcher obtained a random sample of 25 scores. The sample yielded $\bar{x} = 480$ and $s = 95.9$. State the decision rule at the 5% significance level.

(A) Reject H_0 if either $T.S. < -2.064$ or $T.S. > 2.064$.
(B) Reject H_0 if either $T.S. < -1.96$ or $T.S. > 1.96$.
(C) Reject H_0 if $T.S. > 1.645$.
(D) It cannot be determined from the information given.

429. A local deli chain advertises that its "heart-healthy" sandwich contains a mere 10 fat grams. To test the claim, a nutritionist selected a random sample of 4 heart-healthy sandwiches. The sample yielded $\bar{x} = 11$ grams and $s = 1.8$ grams. State the decision rule at the 5% significance level for the appropriate hypothesis test.

(A) Reject H_0 if either $T.S. < -1.96$ or $T.S. > 1.96$.
(B) Reject H_0 if either $T.S. < -2.353$ or $T.S. > 2.353$.
(C) Reject H_0 if either $T.S. < -3.182$ or $T.S. > 3.182$.
(D) It cannot be determined from the information given.

430. A plumbing company has found that the standard deviation for the average length of time, in minutes, for installing a bathtub is 30 minutes. The plumbing company advertises that a bathtub installation takes no more than 2 hours. The owner of the company hired a statistician to test the company's claim. The statistician randomly selected a sample of 40 bathtub installations. The sample yielded $\bar{x} = 150$ minutes and $s = 35$ minutes. State the decision rule at the 10% significance level.

(A) Reject H_0 if $T.S. > 1.303$.
(B) Reject H_0 if $T.S. > 1.282$.
(C) Reject H_0 if $T.S. > 1.645$.
(D) It cannot be determined from the information given.

431. The weights of newborn baby girls born at a county hospital are normally distributed. A hospital administrator conducts a study to determine whether the mean weight of all newborn baby girls born at the hospital is less than 100 ounces. A random sample of size $n = 20$ had a mean weight of 95 ounces and a standard deviation of 7 ounces. Assume σ is unknown. To determine whether, at the 1% significance level, these data indicate that the mean weight of all newborn baby girls born at the county hospital is less than 100 ounces, state the decision rule.

(A) Reject H_0 if $T.S. < -2.33$.
(B) Reject H_0 if $T.S. < -2.539$.
(C) Reject H_0 if $T.S. < -1.328$.
(D) It cannot be determined from the information given.

432. A consumer group complains that the mean lifetime of a certain brand of incandescent light bulbs is no more than 500 hours. A random sample of 5 of these light bulbs had a mean lifetime of 520 hours and standard deviation of 50 hours. State the decision rule, at the 10% significance level, for deciding whether the group's claim should be rejected.

(A) Reject H_0 if $T.S. < -1.533$.
(B) Reject H_0 if $T.S. > 1.533$.
(C) Reject H_0 if $T.S. > 3.747$.
(D) It cannot be determined from the information given.

433. For which one of the following combinations is the stated sample size n large enough, for the given value of p_0, to perform a one-sample proportion z-test?

(A) $n = 500,\ p_0 = 0.05$
(B) $n = 300,\ p_0 = 0.01$
(C) $n = 10,\ p_0 = 0.75$
(D) $n = 2{,}000,\ p_0 = 0.998$

434. For which one of the following combinations is the stated sample size n too small, for the given value of p_0, to perform a one-sample proportion z-test?

(A) $n = 80,\ p_0 = 0.18$
(B) $n = 100,\ p_0 = 0.4$
(C) $n = 10,\ p_0 = 0.5$
(D) $n = 120,\ p_0 = 0.96$

For questions 435–440, state the decision rule for the test of hypothesis in terms of a z-test statistic and appropriate rejection region at the indicated significance level.

435. A random check of 500 lightning cables manufactured by a company found 15 were defective. The company's quality-control guidelines require that the proportion of its lightning cables that are defective must be less than 0.05. The manufacturer wants to determine whether it is meeting its quality-control guidelines. State the decision rule at the 1% significance level.

(A) Reject H_0 if $T.S. > 2.326$.
(B) Reject H_0 if $T.S. > 1.645$.
(C) Reject H_0 if $T.S. < -2.326$.
(D) It cannot be determined from the information given.

436. In a sample for a government study of 80 working women in a metropolitan area, 30 women are over the age of 60. Government officials want to know whether the sample data dispute an earlier claim that the true proportion of working women over the age of 60 in the metropolitan area is no more than 0.18. State the decision rule at the 5% significance level.

(A) Reject H_0 if $T.S. > 1.645$.
(B) Reject H_0 if $T.S. < -1.645$.
(C) Reject H_0 if either $T.S. < -1.96$ or $T.S. > 1.96$.
(D) It cannot be determined from the information given.

437. Suppose a lobbyist claims that the percent of registered voters who prefer a proposed measure is 98%. A pollster plans to conduct a survey of 200 registered voters to test the lobbyist's claim. State the decision rule at the 5% significance level.

(A) Reject H_0 if either $T.S. < -2.326$ or $T.S. > 2.326$.
(B) Reject H_0 if either $T.S. < -1.96$ or $T.S. > 1.96$.
(C) Reject H_0 if either $T.S. < -1.645$ or $T.S. > 1.645$.
(D) It cannot be determined from the information given.

438. The advertising manager for a radio station wants to test the hypothesis that the percent of radio listeners in the area who listen to the manager's radio station is at least 40%. In a random sample of 100 radio listeners in the area, 34 said they listen to the manager's station. State the decision rule at the 5% significance level.

(A) Reject H_0 if $T.S. < -1.645$.
(B) Reject H_0 if $T.S. > 1.645$.
(C) Reject H_0 if $T.S. < -1.96$.
(D) It cannot be determined from the information given.

439. A professor of interior design conjectures that the proportion of assisted-living residents who prefer blue on their bedroom walls is greater than 0.75. In a random sample of 300 assisted-living residents, 240 said they prefer blue on bedroom walls. State the decision rule for whether these data support the conjecture at the 5% significance level.

(A) Reject H_0 if $T.S. > 1.96$.
(B) Reject H_0 if $T.S. > 1.645$.
(C) Reject H_0 if $T.S. < -1.96$.
(D) It cannot be determined from the information given.

440. A market researcher predicts that the percent of households in a particular rural area that have no vehicle is at least 10%. To test this prediction, the researcher's team conducts a random sample of 120 households, yielding 6 households with no vehicle. State the decision rule at the 10% significance level.

(A) Reject H_0 if either $T.S. < -1.282$ or $T.S. > 1.282$.
(B) Reject H_0 if $T.S. > 1.282$.
(C) Reject H_0 if $T.S. < -1.282$.
(D) It cannot be determined from the information given.

For questions 441–442, compute the test statistic for the hypothesis test by formula. (Round your answers, as needed.)

441. A random check of 500 lightning cables manufactured by a company found 15 were defective. The company's quality-control guidelines require that the proportion of lightning cables it manufactures that are defective must be less than 0.05. The manufacturer wants to determine whether it is meeting its quality-control guidelines. Compute *T.S.*

(A) 0.0918
(B) −0.0918
(C) 2.052
(D) −2.052

442. In a sample for a government study of 80 working women in a metropolitan area, 30 were over the age of 60. Government officials want to know whether the sample data dispute an earlier claim that the true proportion of working women over the age of 60 in the metropolitan area is no more than 0.18. Compute *T.S.*

(A) 4.540
(B) 0.195
(C) −0.195
(D) −4.540

For questions 443–445, compute the test statistic for the appropriate hypothesis test by using a graphing calculator. (Round your answers, as needed.)

443. The advertising manager at a radio station wants to test the hypothesis that the percent of radio listeners in the area who listen to the manager's radio station is at least 40%. In a random sample of 100 radio listeners in the area, 34 said they listen to the manager's station. Compute *T.S.*

(A) 1.225
(B) −1.225
(C) 0.110
(D) −0.110

444. A professor of interior design conjectures that the proportion of assisted-living residents who prefer blue for their bedroom walls is greater than 0.75. In a random sample of 300 assisted-living residents, 240 said they prefer blue on bedroom walls. The professor wants to know whether these data support the conjecture. Compute *T.S.*

(A) 2.000
(B) 2.275
(C) −2.000
(D) −2.275

445. A market researcher predicts that at least 10% of households in a particular rural area have no vehicle. To test the prediction, the researcher's team conducted a random sample of 120 households, yielding 6 households with no vehicle. Compute the appropriate *T.S.*

(A) 0.0339
(B) −0.0339
(C) 1.826
(D) −1.826

For questions 446–448, using a graphing calculator, compute the *p*-value for the appropriate hypothesis test, and indicate whether it is significant (*) or not significant (N.S.) at the $\alpha = 0.05$ level. (Round your answers, as needed.)

446. The advertising manager for a radio station wants to test the hypothesis that the percent of radio listeners in the area who listen to the manager's radio station is at least 40%. In a random sample of 100 radio listeners in the area, 34 said they listen to the manager's station. Compute the *p*-value, and indicate its significance.

(A) 0.2207 N.S.
(B) 0.1103 N.S.
(C) 0.2207*
(D) 0.1103*

447. A professor of interior design conjectures that the proportion of assisted-living residents who prefer blue for their bedroom walls is greater than 0.75. In a random sample of 300 assisted-living residents, 240 said they prefer blue on bedroom walls. The professor wants to know whether these data support the conjecture. Compute the *p*-value, and indicate its significance.

(A) 0.0455 N.S.
(B) 0.0455*
(C) 0.0228 N.S.
(D) 0.0228*

448. A market researcher predicts that at least 10% of households in a particular rural area have no vehicle. To test the prediction, the researcher's team conducts a random sample of 120 households, yielding 6 households with no vehicle. Compute the *p*-value, and indicate its significance.

(A) 0.0679*
(B) 0.0679 N.S.
(C) 0.0339*
(D) 0.0339 N.S.

CHAPTER 12

Two-Sample Hypothesis Tests of Means and Proportions

449. A retail store executive wants to know whether the store's average Internet sales exceed its average mail-order sales. A random sample of 15 Internet sales yielded a mean sale amount of $86.40 with a standard deviation of $18.75. A random sample of 10 mail-order sales yielded a mean sale amount of $75.20 with a standard deviation of $14.25. Assume that the population variances are equal and that the two populations from which the samples are drawn are normally distributed. To determine whether the true mean Internet sale exceeds the true mean mail-order sale, which inferential statistical test is most appropriate?

(A) two-sample independent *t*-test
(B) paired *t*-test
(C) two-sample proportion *z*-test
(D) none of the above

450. A researcher conducted a study to determine the effectiveness of a reading intervention. The pre- and post-assessment data for 8 students are shown in the following table. Assume the population of differences is normally distributed. To determine whether the intervention produces higher post-assessment scores, which inferential statistical test is most appropriate?

Student ID Number	1	2	3	4	5	6	7	8
Pre-assessment score	44	55	25	54	63	38	31	34
Post-assessment score	55	68	40	55	75	52	49	48

(A) two-sample independent *t*-test
(B) paired *t*-test
(C) two-sample proportion *z*-test
(D) none of the above

451. A researcher conducts a study to determine whether there is a significant age difference in married couples. Data collected by the researcher from six married couples are shown in the following chart. To test the null hypothesis that the true mean difference between the ages of married couples is zero, which inferential statistical test is most appropriate?

Couple ID Code	A	B	C	D	E	F
Husband's age (in years)	65	48	33	65	44	24
Wife's age (in years)	57	49	29	65	41	23
Difference (in years)	8	−1	4	0	3	1

(A) two-sample independent t-test
(B) paired t-test
(C) two-sample proportion z-test
(D) none of the above

452. Interior designers maintain that blue is the favorite hue of most people. An interior designer believes that adults 25 to 39 years old are less likely to prefer blue than are adults 40 to 59 years old. In a random sample of 430 adults 25 to 39 years old, 292 chose blue as their favorite hue, and in a random sample of 245 adults 40 to 59 years old, 180 chose blue. Assuming the sample sizes are sufficiently large, which inferential statistical test is most appropriate for determining whether the sample data support the interior designer's belief?

(A) two-sample independent t-test
(B) paired t-test
(C) two-sample proportion z-test
(D) none of the above

453. A consumer group conducts a study to compare the average miles per gallon (mpg) achieved in road tests of two popular car brands. A random sample of 5 cars from Brand 1 were tested, yielding $\bar{x}_1 = 29.5$ mpg and $s_1 = 7.4$ mpg. A random sample of 8 cars from Brand 2 were tested, yielding $\bar{x}_2 = 31.6$ mpg and $s_2 = 8.4$ mpg. Assume that the population variances are equal. To test the null hypothesis of no difference between μ_1, the true average miles per gallon for Brand 1, and μ_2, the true average miles per gallon for Brand 2, which inferential statistical test is most appropriate?

(A) two-sample independent t-test
(B) paired t-test
(C) two-sample proportion z-test
(D) none of the above

454. An experiment was performed to determine whether people who go on a low-fat diet for a month lose weight. The data for the experiment are shown in the following table. Assume the population of differences is normally distributed. To determine whether these data provide evidence that the low-fat one-month diet is effective, which inferential statistical test is most appropriate?

Participant ID Code	A	B	C	D	E	F
Starting weight (in pounds)	165	148	210	154	198	145
Ending weight (in pounds)	159	149	195	150	187	142
Difference (in pounds)	6	−1	15	4	11	3

(A) two-sample independent *t*-test
(B) paired *t*-test
(C) two-sample proportion *z*-test
(D) none of the above

455. A health organization asserts that the prevalence of smoking among male adults 18 years or older in the United States exceeds the prevalence of smoking among female adults 18 years or older. In a survey of 1,000 US adults 18 years or older, 129 of the 516 male respondents identified themselves as smokers, and 87 of the 484 female respondents identified themselves as smokers. Assume the sample sizes are sufficiently large. To test whether the data from the survey support the health organization's assertion, which inferential statistical test is most appropriate?

(A) two-sample independent *t*-test
(B) paired *t*-test
(C) two-sample proportion *z*-test
(D) none of the above

456. Random samples of size $n_1 = 40$ and $n_2 = 30$ were drawn from populations 1 and 2, respectively. The samples yielded $\hat{p}_1 = 0.1$ and $\hat{p}_2 = 0.5$. Before collecting the data, a pollster predicted no difference between p_1 and p_2. Which inferential statistical test is most appropriate for determining whether the data dispute the pollster's prediction?

(A) two-sample independent *t*-test
(B) paired *t*-test
(C) two-sample proportion *z*-test
(D) none of the above

457. A consumer group wants to test the performance of two types of gasoline. Miles per gallon were calculated for 5 pairs of matched vehicles that were driven the same distance under the same conditions. In each pair, one vehicle used Gasoline 1, and the other vehicle used Gasoline 2. The data are displayed in the following table. Assume the population of differences is normally distributed. Which inferential statistical test is most appropriate for testing the hypothesis of no difference in the gasoline fuels?

	Pair 1	Pair 2	Pair 3	Pair 4	Pair 5
Gasoline 1	33.0	24.7	37.6	21.4	35.9
Gasoline 2	35.2	28.1	35.0	23.5	32.5
Difference	−2.2	−3.4	2.6	−2.1	3.4

(A) two-sample independent *t*-test
(B) paired *t*-test
(C) two-sample proportion *z*-test
(D) none of the above

458. A retail store executive wants to know whether the store's average Internet sales exceed its average mail-order sales. A random sample of 15 Internet sales yielded a mean sale amount of $86.40 with a standard deviation of $18.75. A random sample of 10 mail-order sales yielded a mean sale amount of $75.20 with a standard deviation of $14.25. Assume the population variances are equal and the two populations from which the samples are drawn are normally distributed. To determine whether the sample data indicate that the true mean Internet sale amount exceeds the true mean mail-order sale amount, let μ_1 be the true mean Internet sale amount and μ_2 be the true mean mail-order sale amount. State the null and alternative hypotheses for the hypothesis test.

(A) $H_0 : \mu_1 \geq \mu_2; H_a : \mu_1 < \mu_2$
(B) $H_0 : \mu_1 < \mu_2; H_a : \mu_1 \geq \mu_2$
(C) $H_0 : \mu_1 \leq \mu_2; H_a : \mu_1 > \mu_2$
(D) $H_0 : \mu_1 > \mu_2; H_a : \mu_1 \leq \mu_2$

459. A researcher conducted a study to determine the effectiveness of a reading intervention. The pre- and post-assessment data for 8 students are shown in the following table. Assume the population of differences is normally distributed. To determine whether the intervention produces higher post-assessment scores, let $\mu_1 - \mu_2 = \mu_D$, where μ_1 is the true mean pre-assessment score and μ_2 is the true mean post-assessment score. State the null and alternative hypotheses for the hypothesis test.

Student ID Number	**1**	**2**	**3**	**4**	**5**	**6**	**7**	**8**
Pre-assessment score	44	55	25	54	63	38	31	34
Post-assessment score	55	68	40	55	75	52	49	48

(A) $H_0: \mu_D \leq 0; H_a: \mu_D > 0$
(B) $H_0: \mu_D < 0; H_a: \mu_D \geq 0$
(C) $H_0: \mu_D > 0; H_a: \mu_D \leq 0$
(D) $H_0: \mu_D \geq 0; H_a: \mu_D < 0$

460. A consumer group conducts a study to compare the average miles per gallon (mpg) achieved in road tests of two popular car brands. A random sample of 5 cars from Brand 1 were tested, yielding $\bar{x}_1 = 29.5$ mpg and $s_1 = 7.4$ mpg. A random sample of 8 cars from Brand 2 were tested, yielding $\bar{x}_2 = 31.6$ mpg and $s_2 = 8.4$ mpg. Assume the population variances are equal and that two populations from which the samples are drawn are normally distributed. To test the null hypothesis of no difference between μ_1, the true average miles per gallon for Brand 1, and μ_2, the true average miles per gallon for Brand 2, state the null and alternative hypotheses for the hypothesis test.

(A) $H_0: \mu_1 \neq \mu_2; H_a: \mu_1 = \mu_2$
(B) $H_0: \mu_1 = \mu_2; H_a: \mu_1 \neq \mu_2$
(C) $H_0: \mu_D \geq 0; H_a: \mu_D < 0$
(D) $H_0: \mu_D \leq 0; H_a: \mu_D > 0$

461. Interior designers maintain that blue is the favorite hue of most people. An interior designer believes that adults 25 to 39 years old are less likely to prefer blue than are adults 40 to 59 years old. In a random sample of 430 adults 25 to 39 years old, 292 chose blue as their favorite hue, and in a random sample of 245 adults 40 to 59 years old, 180 chose blue. To determine whether the sample data support the interior designer's belief, where p_1 is the true proportion of adults 25 to 39 years old whose favorite hue is blue and p_2 is the true proportion of adults 40 to 59 years old whose favorite hue is blue, state the null and alternative hypotheses for the hypothesis test.

(A) $H_0: p_1 \leq p_2; H_a: p_1 > p_2$
(B) $H_0: p_1 < p_2; H_a: p_1 \geq p_2$
(C) $H_0: p_1 \geq p_2; H_a: p_1 < p_2$
(D) $H_0: p_1 > p_2; H_a: p_1 \leq p_2$

462. An experiment was performed to determine whether people who go on a low-fat diet for a month lose weight. The data for the participants in the experiment are shown in the following table. Assume the population of differences is normally distributed, and let $\mu_1 - \mu_2 = \mu_D$, where μ_1 is the true mean starting weight and μ_2 is the true mean ending weight. To determine whether the data provide evidence that the low-fat one-month diet is effective, state the null and alternative hypotheses for the hypothesis test.

Participant ID Code	A	B	C	D	E	F
Starting weight (in pounds)	165	148	210	154	198	145
Ending weight (in pounds)	159	149	195	150	187	142
Difference (in pounds)	6	−1	15	4	11	3

(A) $H_0 : \mu_D \leq 0; H_a : \mu_D > 0$
(B) $H_0 : \mu_D < 0; H_a : \mu_D < 0$
(C) $H_0 : \mu_D = 0; H_a : \mu_D \neq 0$
(D) $H_0 : \mu_D \geq 0; H_a : \mu_D > 0$

463. A health organization asserts that the prevalence of smoking among male adults 18 years or older in the United States exceeds the prevalence of smoking among female adults 18 years or older. In a random survey of 1,000 US adults 18 years or older, 129 of the 516 male respondents identified themselves as smokers, and 87 of the 484 female respondents identified themselves as smokers. Let p_1 be the true proportion of male adults 18 years or older in the United States who are smokers and p_2 be the true proportion of female adults 18 years or older in the United States who are smokers. To determine whether the data from the survey support the health organization's assertion, state the null and alternative hypotheses for the hypothesis test.

(A) $H_0 : p_1 \leq p_2; H_a : p_1 < p_2$
(B) $H_0 : p_1 \leq p_2; H_a : p_1 > p_2$
(C) $H_0 : p_1 = p_2; H_a : p_1 \neq p_2$
(D) $H_0 : p_1 \leq p_2; H_a : p_1 \geq p_2$

464. A consumer group wants to test the performance of two types of gasoline. Miles per gallon were calculated for 5 pairs of matched vehicles that were driven the same distance under the same conditions. In each pair, one vehicle used Gasoline 1, and the other vehicle used Gasoline 2. The data are displayed in the following table. Assume the population of differences is normally distributed, and let $\mu_1 - \mu_2 = \mu_D$, where μ_1 is the true mean miles per gallon for Gasoline 1 and μ_2 is the true mean miles per gallon for Gasoline 2. To test the hypothesis of no difference in the types of gasoline, state the null and alternative hypotheses for the hypothesis test.

	Pair 1	Pair 2	Pair 3	Pair 4	Pair 5
Gasoline 1	33.0	24.7	37.6	21.4	35.9
Gasoline 2	35.2	28.1	35.0	23.5	32.5
Difference	−2.2	−3.4	2.6	−2.1	3.4

(A) $H_0: \mu_D = 0; H_a: \mu_D \neq 0$
(B) $H_0: \mu_D \geq 0; H_a: \mu_D < 0$
(C) $H_0: \mu_D < 0; H_a: \mu_D > 0$
(D) $H_0: \mu_D \neq 0; H_a: \mu_D = 0$

For questions 465–467, compute the test statistic for the appropriate hypothesis test by formula. (Round your answers, as needed.)

465. A retail store executive wants to know whether the store's average Internet sales exceed its average mail-order sales. A random sample of 15 Internet sales yielded a mean sale amount of $86.40 with a standard deviation of $18.75. A random sample of 10 mail-order sales yielded a mean sale amount of $75.20 with a standard deviation of $14.25. Assume the population variances are equal and the two populations from which the samples are drawn are normally distributed. Let μ_1 be the true mean Internet sale amount and μ_2 be the true mean mail-order sale amount. To determine whether the sample data indicate that the true mean Internet sale amount exceeds the true mean mail-order sale amount, compute *T.S.*

(A) −1.536
(B) 1.536
(C) −1.601
(D) 1.601

466. A researcher conducted a study to determine the effectiveness of a reading intervention. The pre- and post-assessment data for 8 students are shown in the following table. Assume the population of differences is normally distributed, and let $\mu_1 - \mu_2 = \mu_D$, where μ_1 is the true mean pre-assessment score and μ_2 is the true mean post-assessment score. To determine whether the intervention produces higher post-assessment scores, compute *T.S.*

Student ID Number	**1**	**2**	**3**	**4**	**5**	**6**	**7**	**8**
Pre-assessment score	44	55	25	54	63	38	31	34
Post-assessment score	55	68	40	55	75	52	49	48

(A) −6.920
(B) 6.473
(C) −6.473
(D) 6.920

467. Interior designers maintain that blue is the favorite hue of most people. An interior designer believes that adults 25 to 39 years old are less likely to prefer blue than are adults 40 to 59 years old. In a random sample of 430 adults 25 to 39 years old, 292 chose blue as their favorite hue, and in a random sample of 245 adults 40 to 59 years old, 180 chose blue. Let p_1 be the true proportion of adults 25 to 39 years old whose favorite hue is blue and p_2 be the true proportion of adults 40 to 59 years old whose favorite hue is blue. To determine whether the survey data support the interior designer's belief, compute *T.S.*

(A) 1.515
(B) 1.555
(C) −1.515
(D) −1.555

For questions 468–471, using a graphing calculator, compute the p-value for the appropriate hypothesis test, and indicate whether it is significant (*) or not significant (N.S.) at the $\alpha = 0.05$ level. (Round your answers, as needed.)

468. A consumer group conducts a study to compare the average miles per gallon (mpg) achieved in road tests of two popular car brands. A random sample of 5 cars from Brand 1 were tested, yielding $\bar{x}_1 = 29.5$ mpg and $s_1 = 7.4$ mpg. A random sample of 8 cars from Brand 2 were tested, yielding $\bar{x}_2 = 31.6$ mpg and $s_2 = 8.4$ mpg. Assume the population variances are equal and the two populations from which the samples are drawn are normally distributed. To test the null hypothesis of no difference in average miles per gallon between the two car brands, compute the p-value and indicate its significance.

(A) 0.6562*
(B) 0.6562 N.S.
(C) 0.6474*
(D) 0.6474 N.S.

469. An experiment was performed to determine whether people who go on a low-fat diet for a month lose weight. The data for the participants in the experiment are shown in the following table. Assume the population of differences is normally distributed, and let $\mu_1 - \mu_2 = \mu_D$, where μ_1 is the true mean before weight and μ_2 is the true mean after weight. To test whether these data provide evidence that the low-fat one-month diet is effective, compute the *p*-value and indicate its significance.

Participant ID Code	A	B	C	D	E	F
Starting weight (in pounds)	165	148	210	154	198	145
Ending weight (in pounds)	159	149	195	150	187	142
Difference (in pounds)	6	−1	15	4	11	3

(A) 0.0129 N.S.
(B) 0.0437 N.S.
(C) 0.0437*
(D) 0.0219*

470. A health organization asserts that the prevalence of smoking among male adults 18 years or older in the United States exceeds the prevalence of smoking among female adults 18 years or older. In a random survey of 1,000 US adults 18 years or older, 129 of the 516 male respondents identified themselves as smokers, and 87 of the 484 female respondents identified themselves as smokers. Let p_1 be the true proportion of male adults 18 years or older in the United States who are smokers and p_2 be the true proportion of female adults 18 years or older in the United States who are smokers. To test whether the data from the survey support the health organization's assertion, compute the *p*-value for the appropriate hypotheses and indicate its significance.

(A) 2.697*
(B) 2.697 N.S.
(C) 0.0035*
(D) 0.0035 N.S.

471. A consumer group wants to test the performance of two types of gasoline. Miles per gallon were calculated for 5 pairs of matched vehicles that were driven the same distance under the same conditions. In each pair, one vehicle used Gasoline 1, and the other vehicle used Gasoline 2. The data are displayed in the following table. Assume the population of differences is normally distributed. To test the hypothesis of no difference in the gasoline fuels, compute the p-value and indicate its significance.

	Pair 1	Pair 2	Pair 3	Pair 4	Pair 5
Gasoline 1	33.0	24.7	37.6	21.4	35.9
Gasoline 2	35.2	28.1	35.0	23.5	32.5
Difference	−2.2	−3.4	2.6	−2.1	3.4

(A) 0.8186*
(B) 0.8186 N.S.
(C) 0.5907*
(D) 0.5907 N.S.

CHAPTER 13

Correlation Analysis and Simple Linear Regression

472. Which one of the following statements is always true?

(A) The greater the value of the correlation coefficient, the stronger is the relationship.

(B) A strong positive correlation between two variables means one of the variables causes the effect of the other variable.

(C) If two variables are independent, their correlation does not exist.

(D) Pearson product-moment correlation coefficients numerically quantify only linear relationships.

473. Which one of the following scatterplots shows the strongest linear relationship between variables X and Y?

(A)

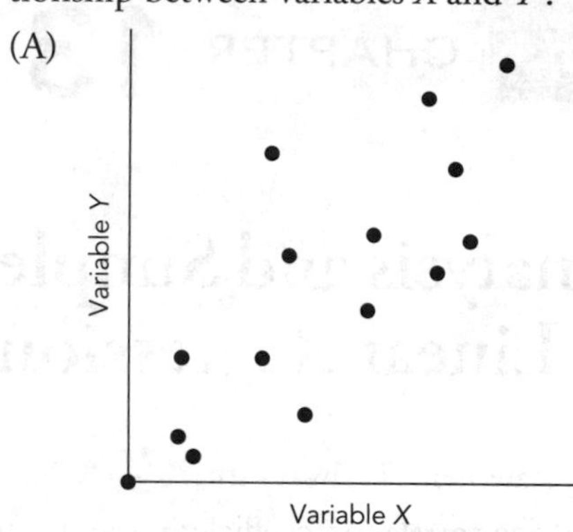

(B)

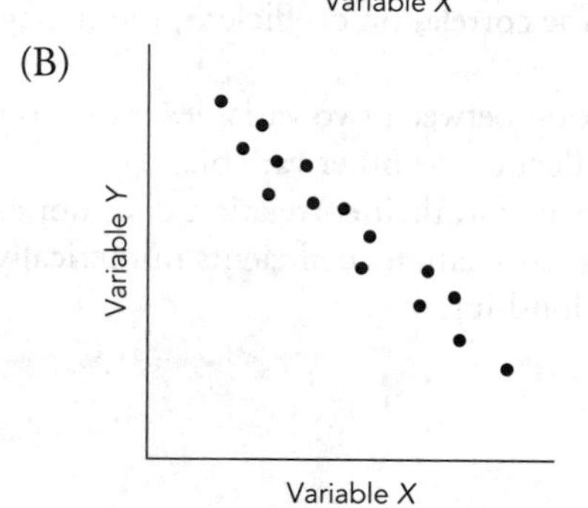

(C)

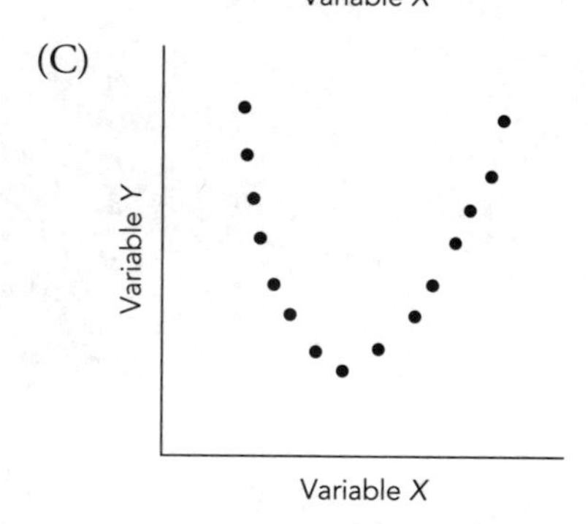

(D)

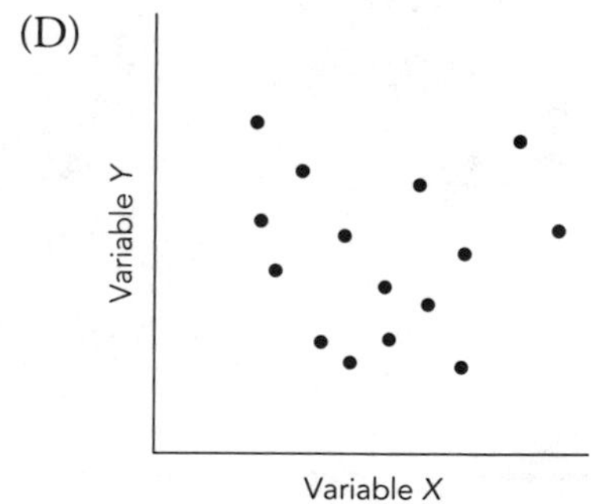

474. Which one of the following statements comparing correlation coefficients is true?

(A) 0.40 is stronger than 0.75.
(B) −0.50 is stronger than −0.95.
(C) 0.82 is weaker than 0.68.
(D) −0.54 is equal in strength to 0.54.

475. The following correlation table shows the correlations between pairs of variables from among the four variables *W*, *X*, *Y*, and *Z*.

Correlation Table for Variables *W*, *X*, *Y*, and *Z*

	Variable W	Variable X	Variable Y	Variable Z
Variable *W*	1			
Variable *X*	−0.85	1		
Variable *Y*	0.47	0.82	1	
Variable *Z*	−0.58	0.60	−0.79	1

The correlation coefficient between which of the following pairs of variables shows the strongest relationship?

(A) *W* and *X*
(B) *Y* and *W*
(C) *X* and *Y*
(D) *Y* and *Z*

476. A biologist is interested in the relationship between age (in years) and the maximum pulse rate (in beats per minute) for women. A sample of five women yielded the following five pairs of data. Calculate the sample product-moment correlation coefficient r by formula. (Round your answer, as needed.)

	Woman 1	Woman 2	Woman 3	Woman 4	Woman 5
Age x (years)	20	39	18	44	50
Pulse rate y (beats per minute)	210	180	200	165	120

(A) $r = -0.904$
(B) $r = -0.749$
(C) $r = 0.904$
(D) $r = 0.749$

477. The following data give the number of hours each of six students in a statistics class spent studying and each student's corresponding score (out of 100 points) on the final exam. Using a graphing calculator, calculate the product-moment correlation coefficient r for the sample. (Round your answer, as needed.)

Student Number	**1**	**2**	**3**	**4**	**5**	**6**
Study time x (in hours)	1.50	2.75	3.00	4.50	5.75	6.00
Score on exam y	54	73	70	82	91	89

(A) $r = -0.969$
(B) $r = -0.939$
(C) $r = 0.939$
(D) $r = 0.969$

478. In a simple linear regression model, which one of the following is NOT an assumption for the error terms?

(A) They are normally distributed.
(B) The variance is zero.
(C) The mean is zero.
(D) The variance is constant.

479. In a simple linear regression model, the estimator for the error variance is

(A) b_0
(B) b_1
(C) SSE
(D) MSE

480. Which one of the following statements is NOT true?

(A) The least-squares regression line passes through the mean point $(\overline{x}, \overline{y})$.
(B) Both correlation and regression analysis assume a linear model.
(C) In a regression analysis, the correlation coefficient and the slope of the regression equation have opposite arithmetic signs.
(D) In a regression analysis, the least-squares procedure finds the line that minimizes the sum of the squared errors from the line.

481. A simple linear regression analysis yields the following results:

$\hat{Y} = -5.2 - 0.75X$

$p = 0.0325$

$r^2 = 0.81$

Which of the following statements best interprets these results?

(A) The correlation between the dependent and independent variables is 0.9.
(B) The correlation between the dependent and independent variables is −0.75.
(C) Eighty-one percent of the time, the regression model will correctly predict values of the dependent variable.
(D) Eighty-one percent of the variation in Y is explained by the regression relationship of Y with X.

482. A least-squares regression line predicting a girl's height using her shoe size as the independent variable is computed from 25 pairs of observations. The girls' heights are measured in centimeters, and the shoe sizes range from 4 to 9. A linear regression model ($p = 0.0435$) yields the following prediction equation: $\hat{Y} = 138.6412 + 4.4723X$. Determine the predicted height (in centimeters) of a girl whose shoe size is 7. (Round your answer, as needed.)

(A) 138.6
(B) 68.7
(C) 169.9
(D) It is not appropriate to use the given prediction equation for a shoe size of 7.

483. A linear regression model ($p = 0.0301$) yields the following prediction equation:

$\hat{Y} = 36.2785 - 1.3041X$

For each one-unit change in the independent variable, what is the estimated change in the mean value of the dependent variable?

(A) −1.3041
(B) 36.2785
(C) −36.2785
(D) 1.3041

484. Data from a sample of 50 paired observations of variables X and Y yield the least-squares regression line $\hat{Y} = 0.17 - 0.56X$ and a sample correlation of -0.70 between the two variables with p-value $= 0.0004$. According to these results, what is the proportion of variance in Y that is statistically explained by the regression equation?

(A) -0.70
(B) 0.49
(C) -0.56
(D) 0.17

For questions 485 and 486, refer to the following information:

Student No.	**1**	**2**	**3**	**4**	**5**	**6**	**7**	**8**	**9**	**10**
Sleep time x	7	9	8	6	4	8	7	6	5	8
GPA y	3.28	3.16	3.75	2.50	2.45	2.91	3.53	3.02	2.68	3.48

The data are from a random sample of 10 students who participated in a study undertaken to investigate the effect of sleep time (measured in average number of hours of sleep per night) on GPA (grade point average, measured on a 4-point scale). A linear regression analysis yields the following results:

```
LinRegTTest
 y=a+bx
 β≠0 and ρ≠0
 t=2.819743049
 p=.022502783
 df=8
↓a=1.696481481
```

```
LinRegTTest
 y=a+bx
 β≠0 and ρ≠0
↑b=.2028703704
 s=.3343766658
 r²=.4984625011
 r=.7060187682
```

485. What is the predicted GPA of a student who averages 10 hours of sleep per night? (Round your answer, as needed.)

(A) 3.73
(B) 3.48
(C) 4.00
(D) It is not appropriate to use the given prediction equation for $x = 10$ hours.

486. What is the predicted GPA of a student who averages 6.5 hours of sleep per night? (Round your answer, as needed.)

(A) 3.02
(B) 3.12
(C) 3.08
(D) It is not appropriate to use the given prediction equation for $x = 6.5$ hours.

For questions 487 and 488, refer to the following information:

A linear regression analysis on a data set of 12 ordered pairs (x, y) yields the following statistics:

$$b_0 = 3.21$$
$$b_1 = 2.46$$
$$\text{SSE} = \sum (y - \hat{y})^2 = 11.815$$
$$\overline{x} = 10.7$$
$$S_{xx} = \sum (x - \overline{x})^2 = 0.9815$$

487. For the model $Y = \beta_0 + \beta_1 + \in$, compute the test statistic for the existence of a linear relationship between X and Y using the following hypotheses:

$$H_0 : \beta_1 = 0$$
$$H_a : \beta_1 \neq 0$$

(A) 2.926
(B) 0.925
(C) 2.242
(D) It cannot be computed from the information given.

488. Compute a 95% confidence interval for β_1.

(A) (0.016, 4.904)
(B) (0.766, 5.654)
(C) (0.046, 4.874)
(D) It cannot be computed from the information given.

CHAPTER 14

Chi-Square Test and ANOVA

489. Which of the following statements is always true?

(A) The chi-square distribution is skewed to the left.
(B) The chi-square distribution is never negative.
(C) The chi-square distribution is symmetric about its mean.
(D) The chi-square distribution can assume values from $-\infty$ to $+\infty$.

490. Use of the chi-square statistics requires that each of the expected cell counts is

(A) at least 10
(B) no more than 10
(C) at least 5
(D) no more than 5

491. In a chi-square test of independence, the test statistic based on a contingency table with 5 rows and 4 columns has how many degrees of freedom?

(A) 9
(B) 15
(C) 12
(D) 7

492. What is the probability that a chi-square random variable with 20 degrees of freedom will exceed 34.170? Use the χ^2 distribution table in Appendix D.

(A) 0.100
(B) 0.050
(C) 0.025
(D) 0.010

493. The following 3×3 contingency table shows the results of classifying 200 students from a random sample of students (50 freshmen, 60 sophomores, 50 juniors, and 40 seniors) from three regional universities according to mathematics anxiety level (low, moderate, and high). If the null hypothesis of no association between level of mathematics anxiety and university classification is true, what is the expected number of seniors who have high mathematics anxiety?

	Anxiety Level			
Classification	**Low**	**Moderate**	**High**	**Total**
Freshmen	15	20	15	50
Sophomores	15	15	30	60
Juniors	20	15	15	50
Seniors	10	10	20	40
Total	60	60	80	200

(A) 24
(B) 20
(C) 8
(D) 16

494. In a chi-square goodness-of-fit test, the null hypothesis is $H_0 : p_1 = p_2 = p_3 = p_4$. The test statistic computed from a sample size of 300 has how many degrees of freedom?

(A) 3
(B) 4
(C) 298
(D) 299

495. The following 2×2 contingency table resulted from a study investigating sleeping habits and job performance at a small factory. One hundred workers were cross-classified according to sleeping habits (poor, good) and job performance (unsatisfactory, satisfactory). If the null hypothesis of no association between sleeping habits and job performance is true, what is the computed chi-square test statistic? (Round your answer, as needed.)

	Job Performance		
Sleeping Habits	**Unsatisfactory**	**Satisfactory**	**Total**
Poor	20	10	30
Good	20	50	70
Total	40	60	100

(A) 12.698
(B) 11.175
(C) 14.080
(D) 6.857

496. Which of the following statements is NOT a required assumption for conducting a one-way ANOVA comparing *k* population means?

(A) The *k* populations under study have equal means.
(B) The *k* populations under study have equal variances.
(C) The *k* populations under study are normally distributed.
(D) Independent random sampling is used to select the samples from the *k* populations.

497. In a two-way ANOVA, the significance of which effect should be examined first?

(A) The significance of the effect of factor 1.
(B) The significance of the effect of factor 2.
(C) The significance of the interaction effect between factors 1 and 2.
(D) It does not matter which effect's significance is examined first.

498. In an ANOVA, which of the following measures is used to find the variability of the observed values of the response variable around their respective treatment means?

(A) SSE
(B) MSE
(C) $SS_{treatment}$
(D) SS_{total}

499. In a one-way ANOVA comparing 4 treatments with 5 observations per treatment, what is the cutoff point for the rejection region associated with the test of hypothesis of equal means? Use the *F*-distribution table in Appendix E.

(A) 5.8025
(B) 3.2389
(C) 8.6923
(D) 3.2874

500. In a one-way ANOVA comparing 4 treatments with 5 observations per treatment, $SS_{treatment} = 763.75$, and SSE = 1,210. Compute the *F*-test statistic, and indicate whether it is significant (*) or not significant (N.S.). (Round your answer to 3 decimal places.)

(A) 3.366 N.S.
(B) 3.366*
(C) 0.631 N.S.
(D) 0.631*

APPENDIX A: BINOMIAL PROBABILITIES (P)

To use the table, first locate n in the leftmost column. Then, for that n, read down to find the row in which the corresponding value of x appears. Now read across that row to the desired value for p.

n	x	p 0.1	0.2	0.3	0.4	0.5	0.6	0.7	0.8	0.9
1	0	0.900	0.800	0.700	0.600	0.500	0.400	0.300	0.200	0.100
	1	0.100	0.200	0.300	0.400	0.500	0.600	0.700	0.800	0.900
2	0	0.810	0.640	0.490	0.360	0.250	0.160	0.090	0.040	0.010
	1	0.180	0.320	0.420	0.480	0.500	0.480	0.420	0.320	0.180
	2	0.010	0.040	0.090	0.160	0.250	0.360	0.490	0.640	0.810
3	0	0.729	0.512	0.343	0.216	0.125	0.064	0.027	0.008	0.001
	1	0.243	0.384	0.441	0.432	0.375	0.288	0.189	0.096	0.027
	2	0.027	0.096	0.189	0.288	0.375	0.432	0.441	0.384	0.243
	3	0.001	0.008	0.027	0.064	0.125	0.216	0.343	0.512	0.729
4	0	0.656	0.410	0.240	0.130	0.062	0.026	0.008	0.002	0.000
	1	0.292	0.410	0.412	0.346	0.250	0.154	0.076	0.026	0.004
	2	0.049	0.154	0.265	0.346	0.375	0.346	0.265	0.154	0.049
	3	0.004	0.026	0.076	0.154	0.250	0.346	0.412	0.410	0.292
	4	0.000	0.002	0.008	0.026	0.062	0.130	0.240	0.410	0.656
5	0	0.590	0.328	0.168	0.078	0.031	0.010	0.002	0.000	0.000
	1	0.328	0.410	0.360	0.259	0.156	0.077	0.028	0.006	0.000
	2	0.073	0.205	0.309	0.346	0.312	0.230	0.132	0.051	0.008
	3	0.008	0.051	0.132	0.230	0.312	0.346	0.309	0.205	0.073
	4	0.000	0.006	0.028	0.077	0.156	0.259	0.360	0.410	0.328
	5	0.000	0.000	0.002	0.010	0.031	0.078	0.168	0.328	0.590
6	0	0.531	0.262	0.118	0.047	0.016	0.004	0.001	0.000	0.000
	1	0.354	0.393	0.303	0.187	0.094	0.037	0.010	0.002	0.000
	2	0.098	0.246	0.324	0.311	0.234	0.138	0.060	0.015	0.001
	3	0.015	0.082	0.185	0.276	0.312	0.276	0.185	0.082	0.015
	4	0.001	0.015	0.060	0.138	0.234	0.311	0.324	0.246	0.098
	5	0.000	0.002	0.010	0.037	0.094	0.187	0.303	0.393	0.354
	6	0.000	0.000	0.001	0.004	0.016	0.047	0.118	0.262	0.531
7	0	0.478	0.210	0.082	0.028	0.008	0.002	0.000	0.000	0.000
	1	0.372	0.367	0.247	0.131	0.055	0.017	0.004	0.000	0.000
	2	0.124	0.275	0.318	0.261	0.164	0.077	0.025	0.004	0.000
	3	0.023	0.115	0.227	0.290	0.273	0.194	0.097	0.029	0.003

		p								
n	x	0.1	0.2	0.3	0.4	0.5	0.6	0.7	0.8	0.9
7	4	0.003	0.029	0.097	0.194	0.273	0.290	0.227	0.115	0.023
	5	0.000	0.004	0.025	0.077	0.164	0.261	0.318	0.275	0.124
	6	0.000	0.000	0.004	0.017	0.055	0.131	0.247	0.367	0.372
	7	0.000	0.000	0.000	0.002	0.008	0.028	0.082	0.210	0.478
8	0	0.430	0.168	0.058	0.017	0.004	0.001	0.000	0.000	0.000
	1	0.383	0.336	0.198	0.090	0.031	0.008	0.001	0.000	0.000
	2	0.149	0.294	0.296	0.209	0.109	0.041	0.010	0.001	0.000
	3	0.033	0.147	0.254	0.279	0.219	0.124	0.047	0.009	0.000
	4	0.005	0.046	0.136	0.232	0.273	0.232	0.136	0.046	0.005
	5	0.000	0.009	0.047	0.124	0.219	0.279	0.254	0.147	0.033
	6	0.000	0.001	0.010	0.041	0.109	0.209	0.296	0.294	0.149
	7	0.000	0.000	0.001	0.008	0.031	0.090	0.198	0.336	0.383
	8	0.000	0.000	0.000	0.001	0.004	0.017	0.058	0.168	0.430
9	0	0.387	0.134	0.040	0.010	0.002	0.000	0.000	0.000	0.000
	1	0.387	0.302	0.156	0.060	0.018	0.004	0.000	0.000	0.000
	2	0.172	0.302	0.267	0.161	0.070	0.021	0.004	0.000	0.000
	3	0.045	0.176	0.267	0.251	0.164	0.074	0.021	0.003	0.000
	4	0.007	0.066	0.172	0.251	0.246	0.167	0.074	0.017	0.001
	5	0.001	0.017	0.074	0.167	0.246	0.251	0.172	0.066	0.007
	6	0.000	0.003	0.021	0.074	0.164	0.251	0.267	0.176	0.045
	7	0.000	0.000	0.004	0.021	0.070	0.161	0.267	0.302	0.172
	8	0.000	0.000	0.000	0.004	0.018	0.060	0.156	0.302	0.387
	9	0.000	0.000	0.000	0.000	0.002	0.010	0.040	0.134	0.387
10	0	0.349	0.107	0.028	0.006	0.001	0.000	0.000	0.000	0.000
	1	0.387	0.268	0.121	0.040	0.010	0.002	0.000	0.000	0.000
	2	0.194	0.302	0.233	0.121	0.044	0.011	0.001	0.000	0.000
	3	0.057	0.201	0.267	0.215	0.117	0.042	0.009	0.001	0.000
	4	0.011	0.088	0.200	0.251	0.205	0.111	0.037	0.006	0.000
	5	0.001	0.026	0.103	0.201	0.246	0.201	0.103	0.026	0.001
	6	0.000	0.006	0.037	0.111	0.205	0.251	0.200	0.088	0.011
	7	0.000	0.001	0.009	0.042	0.117	0.215	0.267	0.201	0.057

(Continued)

		p								
n	*x*	0.1	0.2	0.3	0.4	0.5	0.6	0.7	0.8	0.9
10	8	0.000	0.000	0.001	0.011	0.044	0.121	0.233	0.302	0.194
	9	0.000	0.000	0.000	0.002	0.010	0.040	0.121	0.268	0.387
	10	0.000	0.000	0.000	0.000	0.001	0.006	0.028	0.107	0.349
11	0	0.314	0.086	0.020	0.004	0.000	0.000	0.000	0.000	0.000
	1	0.384	0.236	0.093	0.027	0.005	0.001	0.000	0.000	0.000
	2	0.213	0.295	0.200	0.089	0.027	0.005	0.001	0.000	0.000
	3	0.071	0.221	0.257	0.177	0.081	0.023	0.004	0.000	0.000
	4	0.016	0.111	0.220	0.236	0.161	0.070	0.017	0.002	0.000
	5	0.002	0.039	0.132	0.221	0.226	0.147	0.057	0.010	0.000
	6	0.000	0.010	0.057	0.147	0.226	0.221	0.132	0.039	0.002
	7	0.000	0.002	0.017	0.070	0.161	0.236	0.220	0.111	0.016
	8	0.000	0.000	0.004	0.023	0.081	0.177	0.257	0.221	0.071
	9	0.000	0.000	0.001	0.005	0.027	0.089	0.200	0.295	0.213
	10	0.000	0.000	0.000	0.001	0.005	0.027	0.093	0.236	0.384
	11	0.000	0.000	0.000	0.000	0.000	0.004	0.020	0.086	0.314
12	0	0.282	0.069	0.014	0.002	0.000	0.000	0.000	0.000	0.000
	1	0.377	0.206	0.071	0.017	0.003	0.000	0.000	0.000	0.000
	2	0.230	0.283	0.168	0.064	0.016	0.002	0.000	0.000	0.000
	3	0.085	0.236	0.240	0.142	0.054	0.012	0.001	0.000	0.000
	4	0.021	0.133	0.231	0.213	0.121	0.042	0.008	0.001	0.000
	5	0.004	0.053	0.158	0.227	0.193	0.101	0.029	0.003	0.000
	6	0.000	0.016	0.079	0.177	0.226	0.177	0.079	0.016	0.000
	7	0.000	0.003	0.029	0.101	0.193	0.227	0.158	0.053	0.004
	8	0.000	0.001	0.008	0.042	0.121	0.213	0.231	0.133	0.021
	9	0.000	0.000	0.001	0.012	0.054	0.142	0.240	0.236	0.085
	10	0.000	0.000	0.000	0.002	0.016	0.064	0.168	0.283	0.230
	11	0.000	0.000	0.000	0.000	0.003	0.017	0.071	0.206	0.377
	12	0.000	0.000	0.000	0.000	0.000	0.002	0.014	0.069	0.282
13	0	0.254	0.055	0.001	0.001	0.000	0.000	0.000	0.000	0.000
	1	0.367	0.179	0.054	0.011	0.002	0.000	0.000	0.000	0.000
	2	0.245	0.268	0.139	0.045	0.010	0.001	0.000	0.000	0.000

n	x	0.1	0.2	0.3	0.4	0.5	0.6	0.7	0.8	0.9
13	3	0.100	0.246	0.218	0.111	0.035	0.006	0.001	0.000	0.000
	4	0.028	0.154	0.234	0.184	0.087	0.024	0.003	0.000	0.000
	5	0.006	0.069	0.180	0.221	0.157	0.066	0.014	0.001	0.000
	6	0.001	0.023	0.103	0.197	0.209	0.131	0.044	0.006	0.000
	7	0.000	0.006	0.044	0.131	0.209	0.197	0.103	0.023	0.001
	8	0.000	0.001	0.014	0.066	0.157	0.221	0.180	0.069	0.006
	9	0.000	0.000	0.003	0.024	0.087	0.184	0.234	0.154	0.028
	10	0.000	0.000	0.001	0.006	0.035	0.111	0.218	0.246	0.100
	11	0.000	0.000	0.000	0.001	0.010	0.045	0.139	0.268	0.245
	12	0.000	0.000	0.000	0.000	0.002	0.011	0.054	0.179	0.367
	13	0.000	0.000	0.000	0.000	0.000	0.001	0.010	0.055	0.254
14	0	0.229	0.044	0.007	0.001	0.000	0.000	0.000	0.000	0.000
	1	0.356	0.154	0.041	0.007	0.001	0.000	0.000	0.000	0.000
	2	0.257	0.250	0.113	0.032	0.006	0.001	0.000	0.000	0.000
	3	0.114	0.250	0.194	0.085	0.022	0.003	0.000	0.000	0.000
	4	0.035	0.172	0.229	0.155	0.061	0.014	0.001	0.000	0.000
	5	0.008	0.086	0.196	0.207	0.122	0.041	0.007	0.000	0.000
	6	0.001	0.032	0.126	0.207	0.183	0.092	0.023	0.002	0.000
	7	0.000	0.009	0.062	0.157	0.209	0.157	0.062	0.009	0.000
	8	0.000	0.002	0.023	0.092	0.183	0.207	0.126	0.032	0.001
	9	0.000	0.000	0.007	0.041	0.122	0.207	0.196	0.086	0.008
	10	0.000	0.000	0.001	0.014	0.061	0.155	0.229	0.172	0.035
	11	0.000	0.000	0.000	0.003	0.022	0.085	0.194	0.250	0.114
	12	0.000	0.000	0.000	0.001	0.006	0.032	0.113	0.250	0.257
	13	0.000	0.000	0.000	0.000	0.001	0.007	0.041	0.154	0.356
	14	0.000	0.000	0.000	0.000	0.000	0.001	0.007	0.044	0.229
15	0	0.206	0.035	0.005	0.000	0.000	0.000	0.000	0.000	0.000
	1	0.343	0.132	0.031	0.005	0.000	0.000	0.000	0.000	0.000
	2	0.267	0.231	0.092	0.022	0.003	0.000	0.000	0.000	0.000
	3	0.129	0.250	0.170	0.063	0.014	0.002	0.000	0.000	0.000
	4	0.043	0.188	0.219	0.127	0.042	0.007	0.001	0.000	0.000

(*Continued*)

		p								
n	x	0.1	0.2	0.3	0.4	0.5	0.6	0.7	0.8	0.9
15	5	0.010	0.103	0.206	0.186	0.092	0.024	0.003	0.000	0.000
	6	0.002	0.043	0.147	0.207	0.153	0.061	0.012	0.001	0.000
	7	0.000	0.014	0.081	0.177	0.196	0.118	0.035	0.003	0.000
	8	0.000	0.003	0.035	0.118	0.196	0.177	0.081	0.014	0.000
	9	0.000	0.001	0.017	0.061	0.153	0.207	0.147	0.043	0.002
	10	0.000	0.000	0.003	0.024	0.092	0.186	0.206	0.103	0.010
	11	0.000	0.000	0.001	0.007	0.042	0.127	0.219	0.188	0.043
	12	0.000	0.000	0.000	0.002	0.014	0.063	0.170	0.250	0.129
	13	0.000	0.000	0.000	0.000	0.003	0.022	0.092	0.231	0.267
	14	0.000	0.000	0.000	0.000	0.000	0.005	0.031	0.132	0.343
	15	0.000	0.000	0.000	0.000	0.000	0.000	0.005	0.035	0.206
16	0	0.185	0.028	0.003	0.000	0.000	0.000	0.000	0.000	0.000
	1	0.329	0.113	0.023	0.003	0.000	0.000	0.000	0.000	0.000
	2	0.275	0.211	0.073	0.015	0.002	0.000	0.000	0.000	0.000
	3	0.142	0.246	0.146	0.047	0.009	0.001	0.000	0.000	0.000
	4	0.051	0.200	0.204	0.101	0.028	0.004	0.000	0.000	0.000
	5	0.014	0.120	0.210	0.162	0.067	0.014	0.001	0.000	0.000
	6	0.003	0.055	0.165	0.198	0.122	0.039	0.006	0.000	0.000
	7	0.000	0.020	0.101	0.189	0.175	0.084	0.019	0.001	0.000
	8	0.000	0.006	0.049	0.142	0.196	0.142	0.049	0.006	0.000
	9	0.000	0.001	0.019	0.084	0.175	0.189	0.101	0.020	0.000
	10	0.000	0.000	0.006	0.039	0.122	0.198	0.165	0.055	0.003
	11	0.000	0.000	0.001	0.014	0.067	0.162	0.210	0.120	0.014
	12	0.000	0.000	0.000	0.004	0.028	0.101	0.204	0.200	0.051
	13	0.000	0.000	0.000	0.001	0.009	0.047	0.146	0.246	0.142
	14	0.000	0.000	0.000	0.000	0.002	0.015	0.073	0.211	0.275
	15	0.000	0.000	0.000	0.000	0.000	0.003	0.023	0.113	0.329
	16	0.000	0.000	0.000	0.000	0.000	0.000	0.003	0.028	0.185
17	0	0.167	0.023	0.002	0.000	0.000	0.000	0.000	0.000	0.000
	1	0.315	0.096	0.017	0.002	0.000	0.000	0.000	0.000	0.000
	2	0.280	0.191	0.058	0.010	0.001	0.000	0.000	0.000	0.000

n	x	p 0.1	0.2	0.3	0.4	0.5	0.6	0.7	0.8	0.9
17	3	0.156	0.239	0.125	0.034	0.005	0.000	0.000	0.000	0.000
	4	0.060	0.209	0.187	0.080	0.018	0.002	0.000	0.000	0.000
	5	0.017	0.136	0.208	0.138	0.047	0.008	0.001	0.000	0.000
	6	0.004	0.068	0.178	0.184	0.094	0.024	0.003	0.000	0.000
	7	0.001	0.027	0.120	0.193	0.148	0.057	0.009	0.000	0.000
	8	0.000	0.008	0.064	0.161	0.185	0.107	0.028	0.002	0.000
	9	0.000	0.002	0.028	0.107	0.185	0.161	0.064	0.008	0.000
	10	0.000	0.000	0.009	0.057	0.148	0.193	0.120	0.027	0.001
	11	0.000	0.000	0.003	0.024	0.094	0.184	0.178	0.068	0.004
	12	0.000	0.000	0.001	0.008	0.047	0.138	0.208	0.136	0.017
	13	0.000	0.000	0.000	0.002	0.018	0.080	0.187	0.209	0.060
	14	0.000	0.000	0.000	0.000	0.005	0.034	0.125	0.239	0.156
	15	0.000	0.000	0.000	0.000	0.001	0.010	0.058	0.191	0.280
	16	0.000	0.000	0.000	0.000	0.000	0.002	0.017	0.096	0.315
	17	0.000	0.000	0.000	0.000	0.000	0.000	0.002	0.023	0.167
18	0	0.150	0.018	0.002	0.000	0.000	0.000	0.000	0.000	0.000
	1	0.300	0.081	0.013	0.001	0.000	0.000	0.000	0.000	0.000
	2	0.284	0.172	0.046	0.007	0.001	0.000	0.000	0.000	0.000
	3	0.168	0.230	0.105	0.025	0.003	0.000	0.000	0.000	0.000
	4	0.070	0.215	0.168	0.061	0.012	0.001	0.000	0.000	0.000
	5	0.022	0.151	0.202	0.115	0.033	0.004	0.000	0.000	0.000
	6	0.005	0.082	0.187	0.166	0.071	0.015	0.001	0.000	0.000
	7	0.001	0.035	0.138	0.189	0.121	0.037	0.005	0.000	0.000
	8	0.000	0.012	0.081	0.173	0.167	0.077	0.015	0.001	0.000
	9	0.000	0.003	0.039	0.128	0.185	0.128	0.039	0.003	0.000
	10	0.000	0.001	0.015	0.077	0.167	0.173	0.081	0.012	0.000
	11	0.000	0.000	0.005	0.037	0.121	0.189	0.138	0.035	0.001
	12	0.000	0.000	0.001	0.015	0.071	0.166	0.187	0.082	0.005
	13	0.000	0.000	0.000	0.004	0.033	0.115	0.202	0.151	0.022
	14	0.000	0.000	0.000	0.001	0.012	0.061	0.168	0.215	0.070
	15	0.000	0.000	0.000	0.000	0.003	0.025	0.105	0.230	0.168

(*Continued*)

		p								
n	*x*	0.1	0.2	0.3	0.4	0.5	0.6	0.7	0.8	0.9
18	16	0.000	0.000	0.000	0.000	0.001	0.007	0.046	0.172	0.284
	17	0.000	0.000	0.000	0.000	0.000	0.001	0.013	0.081	0.300
	18	0.000	0.000	0.000	0.000	0.000	0.000	0.002	0.018	0.150
19	0	0.135	0.014	0.001	0.000	0.000	0.000	0.000	0.000	0.000
	1	0.285	0.068	0.009	0.001	0.000	0.000	0.000	0.000	0.000
	2	0.285	0.154	0.036	0.005	0.000	0.000	0.000	0.000	0.000
	3	0.180	0.218	0.087	0.017	0.002	0.000	0.000	0.000	0.000
	4	0.080	0.218	0.149	0.047	0.007	0.001	0.000	0.000	0.000
	5	0.027	0.164	0.192	0.093	0.022	0.002	0.000	0.000	0.000
	6	0.007	0.095	0.192	0.145	0.052	0.008	0.001	0.000	0.000
	7	0.001	0.044	0.153	0.180	0.096	0.024	0.002	0.000	0.000
	8	0.000	0.017	0.098	0.180	0.144	0.053	0.008	0.000	0.000
	9	0.000	0.005	0.051	0.146	0.176	0.098	0.022	0.001	0.000
	10	0.000	0.001	0.022	0.098	0.176	0.146	0.051	0.005	0.000
	11	0.000	0.000	0.008	0.053	0.144	0.180	0.098	0.017	0.000
	12	0.000	0.000	0.002	0.024	0.096	0.180	0.153	0.044	0.001
	13	0.000	0.000	0.001	0.008	0.052	0.145	0.192	0.095	0.007
	14	0.000	0.000	0.000	0.002	0.022	0.093	0.192	0.164	0.027
	15	0.000	0.000	0.000	0.001	0.007	0.047	0.149	0.218	0.080
	16	0.000	0.000	0.000	0.000	0.002	0.017	0.087	0.218	0.180
	17	0.000	0.000	0.000	0.000	0.000	0.005	0.036	0.154	0.285
	18	0.000	0.000	0.000	0.000	0.000	0.001	0.009	0.068	0.285
	19	0.000	0.000	0.000	0.000	0.000	0.000	0.001	0.014	0.135
20	0	0.122	0.012	0.001	0.000	0.000	0.000	0.000	0.000	0.000
	1	0.270	0.058	0.007	0.000	0.000	0.000	0.000	0.000	0.000
	2	0.285	0.137	0.028	0.003	0.000	0.000	0.000	0.000	0.000
	3	0.190	0.205	0.072	0.012	0.001	0.000	0.000	0.000	0.000
	4	0.090	0.218	0.130	0.035	0.005	0.000	0.000	0.000	0.000
	5	0.032	0.175	0.179	0.075	0.015	0.001	0.000	0.000	0.000
	6	0.009	0.109	0.192	0.124	0.037	0.005	0.000	0.000	0.000
	7	0.002	0.055	0.164	0.166	0.074	0.015	0.001	0.000	0.000

		p								
n	*x*	0.1	0.2	0.3	0.4	0.5	0.6	0.7	0.8	0.9
20	8	0.000	0.022	0.114	0.180	0.120	0.035	0.004	0.000	0.000
	9	0.000	0.007	0.065	0.160	0.160	0.071	0.012	0.000	0.000
	10	0.000	0.002	0.031	0.117	0.176	0.117	0.031	0.002	0.000
	11	0.000	0.000	0.012	0.071	0.160	0.160	0.065	0.007	0.000
	12	0.000	0.000	0.004	0.035	0.120	0.180	0.114	0.022	0.000
	13	0.000	0.000	0.001	0.015	0.074	0.166	0.164	0.055	0.002
	14	0.000	0.000	0.000	0.005	0.037	0.124	0.192	0.109	0.009
	15	0.000	0.000	0.000	0.001	0.015	0.075	0.179	0.175	0.032
	16	0.000	0.000	0.000	0.000	0.005	0.035	0.130	0.218	0.090
	17	0.000	0.000	0.000	0.000	0.001	0.012	0.072	0.205	0.190
	18	0.000	0.000	0.000	0.000	0.000	0.003	0.028	0.137	0.285
	19	0.000	0.000	0.000	0.000	0.000	0.000	0.007	0.058	0.270
	20	0.000	0.000	0.000	0.000	0.000	0.000	0.001	0.012	0.122

APPENDIX B: CUMULATIVE STANDARD NORMAL DISTRIBUTION, $-\infty$ to z

Probabilities for Negative Z-Scores

Numerical entries represent the probability that a standard normal random variable is between $-\infty$ and z, where $z = (x - \mu)/\sigma$. Read down the left to find the first two digits in z and then across the column headings for the hundredths place. Notice that the probabilities given in this table represent the area to the LEFT of the z-score.

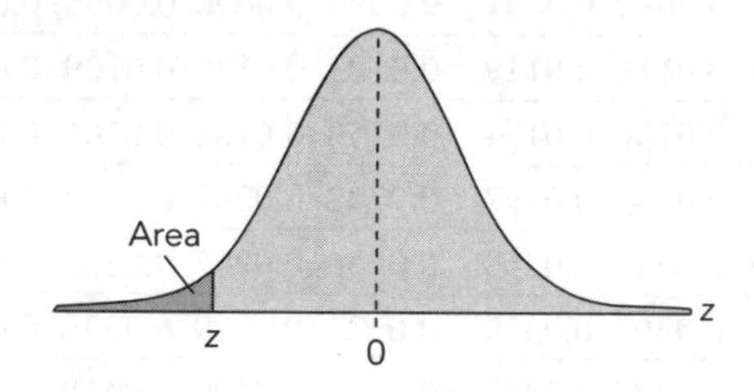

z	0.09	0.08	0.07	0.06	0.05	0.04	0.03	0.02	0.01	0.00
–3.4	0.0002	0.0003	0.0003	0.0003	0.0003	0.0003	0.0003	0.0003	0.0003	0.0003
–3.3	0.0003	0.0004	0.0004	0.0004	0.0004	0.0004	0.0004	0.0005	0.0005	0.0005
–3.2	0.0005	0.0005	0.0005	0.0006	0.0006	0.0006	0.0006	0.0006	0.0007	0.0007
–3.1	0.0007	0.0007	0.0008	0.0008	0.0008	0.0008	0.0009	0.0009	0.0009	0.0010
–3.0	0.0010	0.0010	0.0011	0.0011	0.0011	0.0012	0.0012	0.0013	0.0013	0.0013
–2.9	0.0014	0.0014	0.0015	0.0015	0.0016	0.0016	0.0017	0.0018	0.0018	0.0019
–2.8	0.0019	0.0020	0.0021	0.0021	0.0022	0.0023	0.0023	0.0024	0.0025	0.0026
–2.7	0.0026	0.0027	0.0028	0.0029	0.0030	0.0031	0.0032	0.0033	0.0034	0.0035
–2.6	0.0036	0.0037	0.0038	0.0039	0.0040	0.0041	0.0043	0.0044	0.0045	0.0047
–2.5	0.0048	0.0049	0.0051	0.0052	0.0054	0.0055	0.0057	0.0059	0.0060	0.0062
–2.4	0.0064	0.0066	0.0068	0.0069	0.0071	0.0073	0.0075	0.0078	0.0080	0.0082
–2.3	0.0084	0.0087	0.0089	0.0091	0.0094	0.0096	0.0099	0.0102	0.0104	0.0107
–2.2	0.0110	0.0113	0.0116	0.0119	0.0122	0.0125	0.0129	0.0132	0.0136	0.0139
–2.1	0.0143	0.0146	0.0150	0.0154	0.0158	0.0162	0.0166	0.0170	0.0174	0.0179
–2.0	0.0183	0.0188	0.0192	0.0197	0.0202	0.0207	0.0212	0.0217	0.0222	0.0228
–1.9	0.0233	0.0239	0.0244	0.0250	0.0256	0.0262	0.0268	0.0274	0.0281	0.0287
–1.8	0.0294	0.0301	0.0307	0.0314	0.0322	0.0329	0.0336	0.0344	0.0351	0.0359
–1.7	0.0367	0.0375	0.0384	0.0392	0.0401	0.0409	0.0418	0.0427	0.0436	0.0446
–1.6	0.0455	0.0465	0.0475	0.0485	0.0495	0.0505	0.0516	0.0526	0.0537	0.0548
–1.5	0.0559	0.0571	0.0582	0.0594	0.0606	0.0618	0.0630	0.0643	0.0655	0.0668
–1.4	0.0681	0.0694	0.0708	0.0721	0.0735	0.0749	0.0764	0.0778	0.0793	0.0808
–1.3	0.0823	0.0838	0.0853	0.0869	0.0885	0.0901	0.0918	0.0934	0.0951	0.0968
–1.2	0.0985	0.1003	0.1020	0.1038	0.1056	0.1075	0.1093	0.1112	0.1131	0.1151
–1.1	0.1170	0.1190	0.1210	0.1230	0.1251	0.1271	0.1292	0.1314	0.1335	0.1357
–1.0	0.1379	0.1401	0.1423	0.1446	0.1469	0.1492	0.1515	0.1539	0.1562	0.1587
–0.9	0.1611	0.1635	0.1660	0.1685	0.1711	0.1736	0.1762	0.1788	0.1814	0.1841
–0.8	0.1867	0.1894	0.1922	0.1949	0.1977	0.2005	0.2033	0.2061	0.2090	0.2119
–0.7	0.2148	0.2177	0.2206	0.2236	0.2266	0.2296	0.2327	0.2358	0.2389	0.2420
–0.6	0.2451	0.2483	0.2514	0.2546	0.2578	0.2611	0.2643	0.2676	0.2709	0.2743
–0.5	0.2776	0.2810	0.2843	0.2877	0.2912	0.2946	0.2981	0.3015	0.3050	0.3085
–0.4	0.3121	0.3156	0.3192	0.3228	0.3264	0.3300	0.3336	0.3372	0.3409	0.3446
–0.3	0.3483	0.3520	0.3557	0.3594	0.3632	0.3669	0.3707	0.3745	0.3783	0.3821
–0.2	0.3859	0.3897	0.3936	0.3974	0.4013	0.4052	0.4090	0.4129	0.4168	0.4207
–0.1	0.4247	0.4286	0.4325	0.4364	0.4404	0.4443	0.4483	0.4522	0.4562	0.4602
0.0	0.4641	0.4681	0.4721	0.4761	0.4801	0.4840	0.4880	0.4920	0.4960	0.5000

Probabilities for Positive Z-Scores

Numerical entries represent the probability that a standard normal random variable is between $-\infty$ and z, where $z = (x - \mu)/\sigma$. Read down the left to find the first two digits in z and then across the column headings for the hundredths place. Notice that the probabilities given in this table represent the area to the LEFT of the z-score.

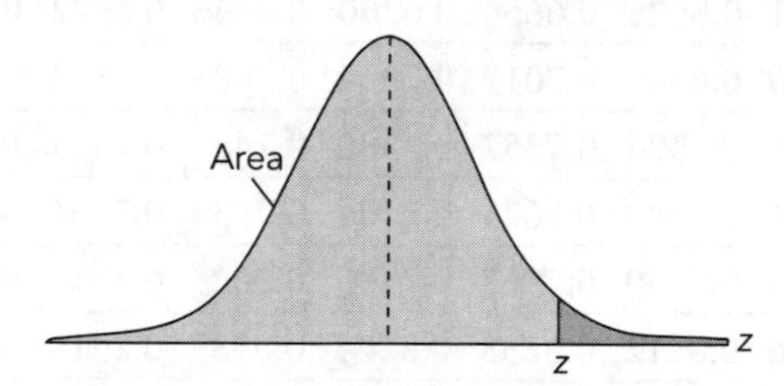

z	0.00	0.01	0.02	0.03	0.04	0.05	0.06	0.07	0.08	0.09
0.0	0.5000	0.5040	0.5080	0.5120	0.5160	0.5199	0.5239	0.5279	0.5319	0.5359
0.1	0.5398	0.5438	0.5478	0.5517	0.5557	0.5596	0.5636	0.5675	0.5714	0.5753
0.2	0.5793	0.5832	0.5871	0.5910	0.5948	0.5987	0.6026	0.6064	0.6103	0.6141
0.3	0.6179	0.6217	0.6255	0.6293	0.6331	0.6368	0.6406	0.6443	0.6480	0.6517
0.4	0.6554	0.6591	0.6628	0.6664	0.6700	0.6736	0.6772	0.6808	0.6844	0.6879
0.5	0.6915	0.6950	0.6985	0.7019	0.7054	0.7088	0.7123	0.7157	0.7190	0.7224
0.6	0.7257	0.7291	0.7324	0.7357	0.7389	0.7422	0.7454	0.7486	0.7517	0.7549
0.7	0.7580	0.7611	0.7642	0.7673	0.7704	0.7734	0.7764	0.7794	0.7823	0.7852
0.8	0.7881	0.7910	0.7939	0.7967	0.7995	0.8023	0.8051	0.8078	0.8106	0.8133
0.9	0.8159	0.8186	0.8212	0.8238	0.8264	0.8289	0.8315	0.8340	0.8365	0.8389
1.0	0.8413	0.8438	0.8461	0.8485	0.8508	0.8531	0.8554	0.8577	0.8599	0.8621
1.1	0.8643	0.8665	0.8686	0.8708	0.8729	0.8749	0.8770	0.8790	0.8810	0.8830
1.2	0.8849	0.8869	0.8888	0.8907	0.8925	0.8944	0.8962	0.8980	0.8997	0.9015
1.3	0.9032	0.9049	0.9066	0.9082	0.9099	0.9115	0.9131	0.9147	0.9162	0.9177
1.4	0.9192	0.9207	0.9222	0.9236	0.9251	0.9265	0.9279	0.9292	0.9306	0.9319
1.5	0.9332	0.9345	0.9357	0.9370	0.9382	0.9394	0.9406	0.9418	0.9429	0.9441
1.6	0.9452	0.9463	0.9474	0.9484	0.9495	0.9505	0.9515	0.9525	0.9535	0.9545
1.7	0.9554	0.9564	0.9573	0.9582	0.9591	0.9599	0.9608	0.9616	0.9625	0.9633
1.8	0.9641	0.9649	0.9656	0.9664	0.9671	0.9678	0.9686	0.9693	0.9699	0.9706
1.9	0.9713	0.9719	0.9726	0.9732	0.9738	0.9744	0.9750	0.9756	0.9761	0.9767
2.0	0.9772	0.9778	0.9783	0.9788	0.9793	0.9798	0.9803	0.9808	0.9812	0.9817
2.1	0.9821	0.9826	0.9830	0.9834	0.9838	0.9842	0.9846	0.9850	0.9854	0.9857
2.2	0.9861	0.9864	0.9868	0.9871	0.9875	0.9878	0.9881	0.9884	0.9887	0.9890
2.3	0.9893	0.9896	0.9898	0.9901	0.9904	0.9906	0.9909	0.9911	0.9913	0.9916
2.4	0.9918	0.9920	0.9922	0.9925	0.9927	0.9929	0.9931	0.9932	0.9934	0.9936
2.5	0.9938	0.9940	0.9941	0.9943	0.9945	0.9946	0.9948	0.9949	0.9951	0.9952
2.6	0.9953	0.9955	0.9956	0.9957	0.9959	0.9960	0.9961	0.9962	0.9963	0.9964
2.7	0.9965	0.9966	0.9967	0.9968	0.9969	0.9970	0.9971	0.9972	0.9973	0.9974
2.8	0.9974	0.9975	0.9976	0.9977	0.9977	0.9978	0.9979	0.9979	0.9980	0.9981
2.9	0.9981	0.9982	0.9982	0.9983	0.9984	0.9984	0.9985	0.9985	0.9986	0.9986
3.0	0.9987	0.9987	0.9987	0.9988	0.9988	0.9989	0.9989	0.9989	0.9990	0.9990
3.1	0.9990	0.9991	0.9991	0.9991	0.9992	0.9992	0.9992	0.9992	0.9993	0.9993
3.2	0.9993	0.9993	0.9994	0.9994	0.9994	0.9994	0.9994	0.9995	0.9995	0.9995
3.3	0.9995	0.9995	0.9995	0.9996	0.9996	0.9996	0.9996	0.9996	0.9996	0.9997
3.4	0.9997	0.9997	0.9997	0.9997	0.9997	0.9997	0.9997	0.9997	0.9997	0.9998

APPENDIX C: CRITICAL VALUES OF t

To use the table, first read down the column labeled "Degrees of Freedom" to locate df $= n - 1$. Then read across that row to the entry in the t_c column under "t_c Area in One Tail" corresponding to the desired value for c.

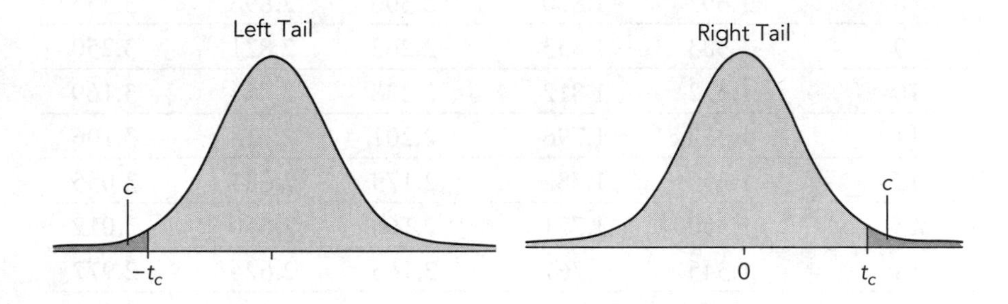

Degrees of Freedom (df)	Area in One Tail				
	$t_{.100}$	$t_{.050}$	$t_{.025}$	$t_{.010}$	$t_{.005}$
1	3.078	6.314	12.706	31.821	63.657
2	1.886	2.920	4.303	6.965	9.925
3	1.638	2.353	3.182	4.541	5.841
4	1.533	2.132	2.776	3.747	4.604
5	1.476	2.015	2.571	3.365	4.032
6	1.440	1.943	2.447	3.143	3.707
7	1.415	1.895	2.365	2.998	3.499
8	1.397	1.860	2.306	2.896	3.355
9	1.383	1.833	2.262	2.821	3.250
10	1.372	1.812	2.228	2.764	3.169
11	1.363	1.796	2.201	2.718	3.106
12	1.356	1.782	2.179	2.681	3.055
13	1.350	1.771	2.160	2.650	3.012
14	1.345	1.761	2.145	2.624	2.977
15	1.341	1.753	2.131	2.602	2.947
16	1.337	1.746	2.120	2.583	2.921
17	1.333	1.740	2.110	2.567	2.898
18	1.330	1.734	2.101	2.552	2.878
19	1.328	1.729	2.093	2.539	2.861
20	1.325	1.725	2.086	2.528	2.845
21	1.323	1.721	2.080	2.518	2.831
22	1.321	1.717	2.074	2.508	2.819
23	1.319	1.714	2.069	2.500	2.807
24	1.318	1.711	2.064	2.492	2.797
25	1.316	1.708	2.060	2.485	2.787
26	1.315	1.706	2.056	2.479	2.779
27	1.314	1.703	2.052	2.473	2.771
28	1.313	1.701	2.048	2.467	2.763
29	1.311	1.699	2.045	2.462	2.756
30	1.310	1.697	2.042	2.457	2.750
31	1.309	1.696	2.040	2.453	2.744

Degrees of Freedom (df)	Area in One Tail				
	$t_{.100}$	$t_{.050}$	$t_{.025}$	$t_{.010}$	$t_{.005}$
32	1.309	1.694	2.037	2.449	2.738
34	1.307	1.691	2.032	2.441	2.728
36	1.306	1.688	2.028	2.434	2.719
38	1.304	1.686	2.024	2.429	2.712
40	1.303	1.684	2.021	2.423	2.704
45	1.301	1.679	2.014	2.412	2.690
50	1.299	1.676	2.009	2.403	2.678
55	1.297	1.673	2.004	2.396	2.668
60	1.296	1.671	2.000	2.390	2.660
65	1.295	1.669	1.997	2.385	2.654
70	1.294	1.667	1.994	2.381	2.648
75	1.293	1.665	1.992	2.377	2.643
80	1.292	1.664	1.990	2.374	2.639
90	1.291	1.662	1.987	2.368	2.632
100	1.290	1.660	1.984	2.364	2.656
120	1.289	1.658	1.980	2.358	2.617
200	1.286	1.653	1.972	2.345	2.601
300	1.284	1.650	1.968	2.339	2.592
400	1.284	1.649	1.966	2.336	2.588
500	1.283	1.648	1.965	2.334	2.586
750	1.283	1.647	1.963	2.331	2.582
1000	1.282	1.646	1.962	2.330	2.581
2000	1.282	1.646	1.961	2.328	2.578
∞	1.282	1.645	1.960	2.326	2.576

APPENDIX D: CRITICAL VALUES OF χ^2

To use the table, first read down the column labeled "Degrees of Freedom" to locate df for the χ^2 test. Then read across that row to the entry in the χ^2_α column corresponding to the desired value for α.

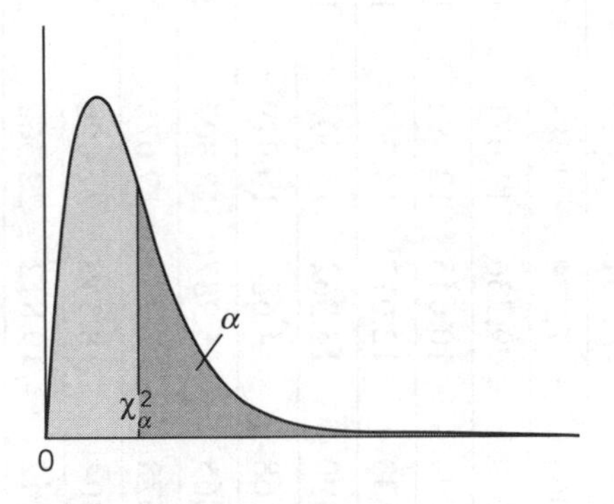

Degrees of Freedom (df)	Area to the Right of the Critical Value									
	$\chi^2_{.995}$	$\chi^2_{.990}$	$\chi^2_{.975}$	$\chi^2_{.950}$	$\chi^2_{.900}$	$\chi^2_{.100}$	$\chi^2_{.050}$	$\chi^2_{.025}$	$\chi^2_{.010}$	$\chi^2_{.005}$
1	—	—	0.001	0.004	0.016	2.706	3.841	5.024	6.635	7.879
2	0.010	0.020	0.051	0.103	0.211	4.605	5.991	7.378	9.210	10.597
3	0.072	0.115	0.216	0.352	0.584	6.251	7.815	9.348	11.345	12.838
4	0.207	0.297	0.484	0.711	1.064	7.779	9.488	11.143	13.277	14.860
5	0.412	0.554	0.831	1.145	1.610	9.236	11.071	12.833	15.086	16.750
6	0.676	0.872	1.237	1.635	2.204	10.645	12.592	14.449	16.812	18.548
7	0.989	1.239	1.690	2.167	2.833	12.017	14.067	16.013	18.475	20.278
8	1.344	1.646	2.180	2.733	3.490	13.362	15.507	17.535	20.090	21.955
9	1.735	2.088	2.700	3.325	4.168	14.684	16.919	19.023	21.666	23.589
10	2.156	2.558	3.247	3.940	4.865	15.987	18.307	20.483	23.209	25.188
11	2.603	3.053	3.816	4.575	5.578	17.275	19.675	21.920	24.725	26.757
12	3.074	3.571	4.404	5.226	6.304	18.549	21.026	23.337	26.217	28.299
13	3.565	4.107	5.009	5.892	7.042	19.812	22.362	24.736	27.688	29.819
14	4.075	4.660	5.629	6.571	7.790	21.064	23.685	26.119	29.141	31.319
15	4.601	5.229	6.262	7.261	8.547	22.307	24.996	27.488	30.578	32.801
16	5.142	5.812	6.908	7.962	9.312	23.542	26.296	28.845	32.000	34.267
17	5.697	6.408	7.564	8.672	10.085	24.769	27.587	30.191	33.409	35.718
18	6.265	7.015	8.231	9.390	10.865	25.989	28.869	31.526	34.805	37.156

19	6.844	7.633	8.907	10.117	11.651	27.204	30.144	32.852	36.191	38.582
20	7.434	8.260	9.591	10.851	12.443	28.412	31.410	34.170	37.566	39.997
21	8.034	8.897	10.283	11.591	13.240	29.615	32.671	35.479	38.932	41.401
22	8.643	9.542	10.982	12.338	14.042	30.813	33.924	36.781	40.289	42.796
23	9.260	10.196	11.689	13.091	14.848	32.007	35.172	38.076	41.638	44.181
24	9.886	10.856	12.401	13.848	15.659	33.196	36.415	39.364	42.980	45.559
25	10.520	11.524	13.120	14.611	16.473	34.382	37.652	40.646	44.314	46.928
26	11.160	12.198	13.844	15.379	17.292	35.563	38.885	41.923	45.642	48.290
27	11.808	12.879	14.573	16.151	18.114	36.741	40.113	43.194	46.963	49.645
28	12.461	13.565	15.308	16.928	18.939	37.916	41.337	44.461	48.278	50.993
29	13.121	14.257	16.047	17.708	19.768	39.087	42.557	45.722	49.588	52.336
30	13.787	14.954	16.791	18.493	20.599	40.256	43.773	46.979	50.892	53.672
40	20.707	22.164	24.433	26.509	29.051	51.805	55.758	59.342	63.691	66.766
50	27.991	29.707	32.357	34.764	37.689	63.167	67.505	71.420	76.154	79.490
60	35.534	37.485	40.482	43.188	46.459	74.397	79.082	83.298	88.379	91.952
70	43.275	45.442	48.758	51.739	55.329	85.527	90.531	95.023	100.425	104.215
80	51.172	53.540	57.153	60.391	64.278	96.578	101.879	106.629	112.329	116.321
90	59.196	61.754	65.647	69.126	73.291	107.565	113.145	118.136	124.116	128.299
100	67.328	70.065	74.222	77.929	82.358	118.498	124.342	129.561	135.807	140.169

APPENDIX E: CRITICAL VALUES OF THE F-DISTRIBUTION ($\alpha = 0.050$)

To use the table, first read across the row labeled "Numerator Degrees of Freedom" to the df for F's numerator. Then find the entry in the intersection of that column with the row under "Denominator Degrees of Freedom" corresponding to F's denominator.

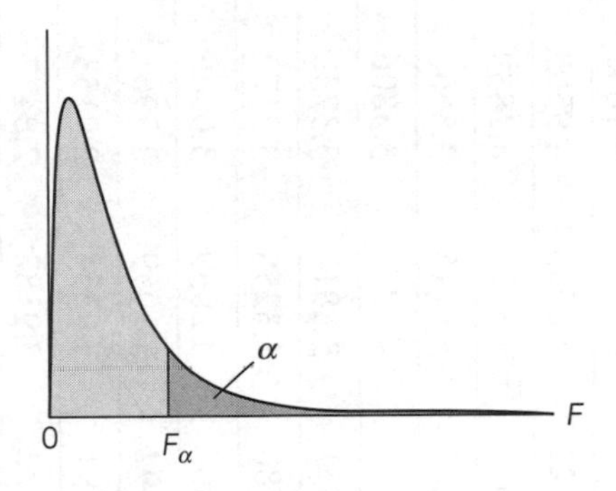

Denominator Degrees of Freedom	Numerator Degrees of Freedom								
	1	2	3	4	5	6	7	8	9
1	161.4476	199.5000	215.7073	224.5832	230.1619	233.9860	236.7684	238.8827	240.5433
2	18.5128	19.0000	19.1643	19.2468	19.2964	19.3295	19.3532	19.3710	19.3848
3	10.1280	9.5521	9.2766	9.1172	9.0135	8.9406	8.8867	8.8452	8.8123
4	7.7086	6.9443	6.5914	6.3882	6.2561	6.1631	6.0942	6.0410	5.9988
5	6.6079	5.7861	5.4095	5.1922	5.0503	4.9503	4.8759	4.8183	4.7725
6	5.9874	5.1433	4.7571	4.5337	4.3874	4.2839	4.2067	4.1468	4.0990
7	5.5914	4.7374	4.3468	4.1203	3.9715	3.8660	3.7870	3.7257	3.6767
8	5.3177	4.4590	4.0662	3.8379	3.6875	3.5806	3.5005	3.4381	3.3881
9	5.1174	4.2565	3.8625	3.6331	3.4817	3.3738	3.2927	3.2296	3.1789
10	4.9646	4.1028	3.7083	3.4780	3.3258	3.2172	3.1355	3.0717	3.0204
11	4.8443	3.9823	3.5874	3.3567	3.2039	3.0946	3.0123	2.9480	2.8962
12	4.7472	3.8853	3.4903	3.2592	3.1059	2.9961	2.9134	2.8486	2.7964
13	4.6672	3.8056	3.4105	3.1791	3.0254	2.9153	2.8321	2.7669	2.7144
14	4.6001	3.7389	3.3439	3.1122	2.9582	2.8477	2.7642	2.6987	2.6458
15	4.5431	3.6823	3.2874	3.0556	2.9013	2.7905	2.7066	2.6408	2.5876
16	4.4940	3.6337	3.2389	3.0069	2.8524	2.7413	2.6572	2.5911	2.5377
17	4.4513	3.5915	3.1968	2.9647	2.8100	2.6987	2.6143	2.5480	2.4943
18	4.4139	3.5546	3.1599	2.9277	2.7729	2.6613	2.5767	2.5102	2.4563
19	4.3807	3.5219	3.1274	2.8951	2.7401	2.6283	2.5435	2.4768	2.4227

	20	4.3512	3.4928	3.0984	2.8661	2.7109	2.5990	2.5140	2.4471	2.3928
	21	4.3248	3.4668	3.0725	2.8401	2.6848	2.5727	2.4876	2.4205	2.3660
	22	4.3009	3.4434	3.0491	2.8167	2.6613	2.5491	2.4638	2.3965	2.3419
	23	4.2793	3.4221	3.0280	2.7955	2.6400	2.5277	2.4422	2.3748	2.3201
	24	4.2597	3.4028	3.0088	2.7763	2.6207	2.5082	2.4226	2.3551	2.3002
	25	4.2417	3.3852	2.9912	2.7587	2.6030	2.4904	2.4047	2.3371	2.2821
	26	4.2252	3.3690	2.9752	2.7426	2.5868	2.4741	2.3883	2.3205	2.2655
	27	4.2100	3.3541	2.9604	2.7278	2.5719	2.4591	2.3732	2.3053	2.2501
	28	4.1960	3.3404	2.9467	2.7141	2.5581	2.4453	2.3593	2.2913	2.2360
	29	4.1830	3.3277	2.9340	2.7014	2.5454	2.4324	2.3463	2.2783	2.2229
	30	4.1709	3.3158	2.9223	2.6896	2.5336	2.4205	2.3343	2.2662	2.2107
	40	4.0847	3.2317	2.8387	2.6060	2.4495	2.3359	2.2490	2.1802	2.1240
	60	4.0012	3.1504	2.7581	2.5252	2.3683	2.2541	2.1665	2.0970	2.0401
	120	3.9201	3.0718	2.6802	2.4472	2.2899	2.1750	2.0868	2.0164	1.9588
	∞	3.8415	2.9957	2.6049	2.3719	2.2141	2.0986	2.0096	1.9384	1.8799

(Continued)

Denominator Degrees of Freedom	Numerator Degrees of Freedom								
	10	11	12	13	14	15	16	17	18
1	241.8817	242.9835	243.9060	244.6898	245.3640	245.9499	246.4639	246.9184	247.3232
2	19.3959	19.4050	19.4125	19.4189	19.4244	19.4291	19.4333	19.4370	19.4402
3	8.7855	8.7633	8.7446	8.7287	8.7149	8.7029	8.6923	8.6829	8.6745
4	5.9644	5.9358	5.9117	5.8911	5.8733	5.8578	5.8441	5.8320	5.8211
5	4.7351	4.7040	4.6777	4.6552	4.6358	4.6188	4.6038	4.5904	4.5785
6	4.0600	4.0274	3.9999	3.9764	3.9559	3.9381	3.9223	3.9083	3.8957
7	3.6365	3.6030	3.5747	3.5503	3.5292	3.5107	3.4944	3.4799	3.4669
8	3.3472	3.3130	3.2839	3.2590	3.2374	3.2184	3.2016	3.1867	3.1733
9	3.1373	3.1025	3.0729	3.0475	3.0255	3.0061	2.9890	2.9737	2.9600
10	2.9782	2.9430	2.9130	2.8872	2.8647	2.8450	2.8276	2.8120	2.7980
11	2.8536	2.8179	2.7876	2.7614	2.7386	2.7186	2.7009	2.6851	2.6709
12	2.7534	2.7173	2.6866	2.6602	2.6371	2.6169	2.5989	2.5828	2.5684
13	2.6710	2.6347	2.6037	2.5769	2.5536	2.5331	2.5149	2.4987	2.4841
14	2.6022	2.5655	2.5342	2.5073	2.4837	2.4630	2.4446	2.4282	2.4134
15	2.5437	2.5068	2.4753	2.4481	2.4244	2.4034	2.3849	2.3683	2.3533
16	2.4935	2.4564	2.4247	2.3973	2.3733	2.3522	2.3335	2.3167	2.3016
17	2.4499	2.4126	2.3807	2.3531	2.3290	2.3077	2.2888	2.2719	2.2567
18	2.4117	2.3742	2.3421	2.3143	2.2900	2.2686	2.2496	2.2325	2.2172
19	2.3779	2.3402	2.3080	2.2800	2.2556	2.2341	2.2149	2.1977	2.1823

	20	2.3479	2.3100	2.2776	2.2495	2.2250	2.2033	2.1840	2.1667	2.1511
	21	2.3210	2.2829	2.2504	2.2222	2.1975	2.1757	2.1563	2.1389	2.1232
	22	2.2967	2.2585	2.2258	2.1975	2.1727	2.1508	2.1313	2.1138	2.0980
	23	2.2747	2.2364	2.2036	2.1752	2.1502	2.1282	2.1086	2.0910	2.0751
	24	2.2547	2.2163	2.1834	2.1548	2.1298	2.1077	2.0880	2.0703	2.0543
	25	2.2365	2.1979	2.1649	2.1362	2.1111	2.0889	2.0691	2.0513	2.0353
	26	2.2197	2.1811	2.1479	2.1192	2.0939	2.0716	2.0518	2.0339	2.0178
	27	2.2043	2.1655	2.1323	2.1035	2.0781	2.0558	2.0358	2.0179	2.0017
	28	2.1900	2.1512	2.1179	2.0889	2.0635	2.0411	2.0210	2.0030	1.9868
	29	2.1768	2.1379	2.1045	2.0755	2.0500	2.0275	2.0073	1.9893	1.9730
	30	2.1646	2.1256	2.0921	2.0630	2.0374	2.0148	1.9946	1.9765	1.9601
	40	2.0772	2.0376	2.0035	1.9738	1.9476	1.9245	1.9037	1.8851	1.8682
	60	1.9926	1.9522	1.9174	1.8870	1.8602	1.8364	1.8151	1.7959	1.7784
	120	1.9105	1.8693	1.8337	1.8026	1.7750	1.7505	1.7285	1.7085	1.6904
	∞	1.8307	1.7886	1.7522	1.7202	1.6918	1.6664	1.6435	1.6228	1.6039

(*Continued*)

Denominator Degrees of Freedom	Numerator Degrees of Freedom						
	19	20	24	30	40	60	120
1	247.6861	248.0131	249.0518	250.0951	251.1432	252.1957	253.2529
2	19.4431	19.4458	19.4541	19.4624	19.4707	19.4791	19.4874
3	8.6670	8.6602	8.6385	8.6166	8.5944	8.5720	8.5494
4	5.8114	5.8025	5.7744	5.7459	5.7170	5.6877	5.6581
5	4.5678	4.5581	4.5272	4.4957	4.4638	4.4314	4.3985
6	3.8844	3.8742	3.8415	3.8082	3.7743	3.7398	3.7047
7	3.4551	3.4445	3.4105	3.3758	3.3404	3.3043	3.2674
8	3.1613	3.1503	3.1152	3.0794	3.0428	3.0053	2.9669
9	2.9477	2.9365	2.9005	2.8637	2.8259	2.7872	2.7475
10	2.7854	2.7740	2.7372	2.6996	2.6609	2.6211	2.5801
11	2.6581	2.6464	2.6090	2.5705	2.5309	2.4901	2.4480
12	2.5554	2.5436	2.5055	2.4663	2.4259	2.3842	2.3410
13	2.4709	2.4589	2.4202	2.3803	2.3392	2.2966	2.2524
14	2.4000	2.3879	2.3487	2.3082	2.2664	2.2229	2.1778
15	2.3398	2.3275	2.2878	2.2468	2.2043	2.1601	2.1141
16	2.2880	2.2756	2.2354	2.1938	2.1507	2.1058	2.0589
17	2.2429	2.2304	2.1898	2.1477	2.1040	2.0584	2.0107
18	2.2033	2.1906	2.1497	2.1071	2.0629	2.0166	1.9681
19	2.1683	2.1555	2.1141	2.0712	2.0264	1.9795	1.9302

	20	2.1370	2.1242	2.0825	2.0391	1.9938	1.9464	1.8963
	21	2.1090	2.0960	2.0540	2.0102	1.9645	1.9165	1.8657
	22	2.0837	2.0707	2.0283	1.9842	1.9380	1.8894	1.8380
	23	2.0608	2.0476	2.0050	1.9605	1.9139	1.8648	1.8128
	24	2.0399	2.0267	1.9838	1.9390	1.8920	1.8424	1.7896
	25	2.0207	2.0075	1.9643	1.9192	1.8718	1.8217	1.7684
	26	2.0032	1.9898	1.9464	1.9010	1.8533	1.8027	1.7488
	27	1.9870	1.9736	1.9299	1.8842	1.8361	1.7851	1.7306
	28	1.9720	1.9586	1.9147	1.8687	1.8203	1.7689	1.7138
	29	1.9581	1.9446	1.9005	1.8543	1.8055	1.7537	1.6981
	30	1.9452	1.9317	1.8874	1.8409	1.7918	1.7396	1.6835
	40	1.8529	1.8389	1.7929	1.7444	1.6928	1.6373	1.5766
	60	1.7625	1.7480	1.7001	1.6491	1.5943	1.5343	1.4673
	120	1.6739	1.6587	1.6084	1.5543	1.4952	1.4290	1.3519
	∞	1.5865	1.5705	1.5173	1.4591	1.3940	1.3180	1.2214

ANSWERS

To assist you with technology, some questions are worked out using the TI-84 Plus Silver Edition calculator. Most graphing calculators have features similar to those on the TI-83/84 series. Learning to use your calculator correctly as a useful tool will greatly enhance your performance in statistics.

Chapter 1: Classifying Data

1. (C) Quantitative data consist of numerical measurements or counts. Lengths are measurements, so these are quantitative data. Choices A, B, and D are qualitative data.

2. (D) Weights are measurements, so these are quantitative data. Choices A, B, and C are qualitative data.

3. (C) Survey responses are neither measurements nor counts, so these are not quantitative data. Freezing temperatures (A) and heights (B) are measurements, so these are quantitative data. Number of people (D) is count data, so these data are quantitative.

4. (A) Qualitative data consist of labels or descriptions of traits (also known as categorical data). Eye colors are descriptions, so these are qualitative data. Choices B, C, and D are quantitative data.

5. (B) Names are labels, so these are qualitative data. Choices A, C, and D are quantitative data.

6. (C) Heights are measurements, so these are quantitative data, not qualitative data. Social security numbers (A) and numbers on football jerseys (B) are numeric values used as labels (they are not a measurement or count), so these are qualitative data. Customer ratings (D) are descriptions, so these are qualitative data.

7. (A) Eye colors are nominal level. When the highest level of a data set is nominal level, the data consist of categories (names, labels, etc.) only. Standard mathematical operations (addition, subtraction, multiplication, division) are not defined when applied to data that is strictly nominal level.

8. (D) Distances are ratio level. When the highest level of a data set is ratio level, the data have a meaningful zero, and arithmetic sums, differences, products, and ratios are meaningful.

9. (D) Number of students is count data, which is ratio level.

10. (B) Survey responses are ordinal level. When the highest level of a data set is ordinal level, the data consist of categories that can be arranged in some meaningful order, but differences between data values are meaningless. Standard mathematical operations (addition, subtraction, multiplication, division) are not defined when applied to data that reach only ordinal level.

11. (B) Nominal (I) and ordinal (II) data are considered qualitative. Interval (III) and ratio (IV) are considered quantitative.

12. (D) Interval (III) and ratio (IV) data are considered quantitative. Nominal (I) and ordinal (II) data are considered qualitative.

13. (A) The number of brown-eyed children is discrete data, because these data are countable. Choices B and C are continuous data. Choice D is neither discrete nor continuous data because these are not quantitative.

14. (A) The number in favor of school uniforms is discrete data, because these data are countable. Choices B, C, and D are continuous data.

15. (C) Heights of plants (in centimeters) are continuous data because these data are measurements that can take on any value within a range of values on a numerical scale. Choices A, B, and D are discrete data.

16. (C) Lengths (in meters) are continuous data because these data are measurements that can take on any value within a range of values on a numerical scale. Choice B is discrete data. Choices A and D are neither discrete nor continuous data because these are not quantitative.

Chapter 2: Organizing Data

17. (C) A circle graph is made by dividing the 360 degrees of the circle that makes the graph into portions that correspond to the proportion for each category. The central angle that should be used to represent the category Comedy/romantic comedy is:

$$\frac{76}{240}(360°) = 114°$$

18. (A) The percent budgeted for rent is 40%. The percent budgeted for food and clothing combined is 17% + 17% = 34%. The difference is 40% − 34% = 6%. Thus, 6% of \$3,500 = 0.06(\$3,500) = \$210 is how much more money is budgeted for rent than for food and clothing combined. Rather than working with the percentages first, you can compute the money amounts first, and then subtract the amounts budgeted for food and clothing from the amount budgeted for rent to obtain the same answer.

19. (B) To read a pictograph, you count the number of symbols shown and then multiply by the number each symbol represents. According to the graph shown, each cat picture represents 2 cat owners. Thus, the number of cat owners who responded "yes" to the question "Do you own a dog?" is $(6)(2) = 12$.

20. (C) According to the bar graph shown, 8 students achieved a grade of D, 6 students achieved a C, 19 students achieved a B, and 3 students achieved an A, for a total of $8 + 6 + 19 + 3 = 36$ students who passed the chemistry course.

21. (C) The lowest profit (revenues – expenses) occurs when the two line graphs are closest together. According to the double line graph, this situation occurred in May. According to the graph, May revenues were $150,000, and expenses were $130,000. Thus, the lowest monthly profit was $150,000 – $130,000 = $20,000.

22. (D) A stem-and-leaf plot (also called stemplot) is a graphical display of data in which each data value is separated into two parts: a stem and a leaf. For a given data value, the leaf gives the rightmost digit, and the stem gives the remaining digits. According to the stem-and-leaf plot shown, the oldest age at which a US president was inaugurated is 69, and the youngest age is 42, giving a difference of $69 - 42 = 27$.

23. (B) According to the stem-and-leaf plot shown, the number of US presidents who were 64 or older at inauguration is 5.

24. (C) According to the stem-and-leaf plot shown, 145 is the starting weight that occurs most often.

25. (D) According to the stem-and-leaf plot shown, 21 of the 36 female participants have a starting weight of at least 145 pounds. Thus, $\frac{21}{36} \approx 0.58 = 58\%$ have a starting weight of at least 145 pounds.

26. (B) According to the dot plot shown, 12 customers waited less than 15 minutes in line.

27. (C) According to the dot plot shown, the greatest amount of time in minutes waited in line by a customer is 19.

28. (B) The class width of a frequency distribution is the upper class limit minus the lower class limit. For the frequency distribution shown, class width $= 799.99 - 500.00 = 1{,}099.99 - 800.00 = 1{,}399.99 - 1{,}100.00 = 1{,}699.99 - 1{,}400.00 = 1{,}999.99 - 1{,}700.00 = 299.99$.

29. (B) The class with the highest frequency is 800.00–1,099.99. The midpoint of this class is:

$$\frac{800.00 + 1{,}099.99}{2} = 949.995$$

30. (D) Based on the frequency histogram shown, about 35 students consume between 0 and 5 soft drinks per week, 50 consume between 6 and 10, and about 25 consume between 11 and 15, for a total of 110 students who consume fewer than 16 soft drinks per week.

31. (C) The majority of the data fall on the left of the distribution with the right side extending out to the right. This shape is described as positively skewed.

32. (C) According to the stacked bar chart shown, Store 3 sold 20 units (20% of 100) of product A, 10 units (10% of 100) of product B, 40 units (40% of 100) of product C, and 30 units (30% of 100) of product D, for a total of 100 units of the 400 units sold. Thus, Store 3 sold $\frac{100}{400} = 25\%$ of the 400 total units sold by the four stores.

Chapter 3: Measures of Central Tendency

33. (C) The mean is $\frac{\sum x}{n} = \frac{280}{10} = 28$. The mean of a data set is the arithmetic average of the data values. Symbolically, the mean is equal to $\frac{\sum x}{n}$, where $\sum x$ represents the sum of all the data values and n represents the number of data values in the data set.

34. (C) Mean $= \frac{\sum x}{n} = \frac{51}{6} = 8.5$

35. (A) Mean $= \frac{\sum x}{n} = \frac{41.8}{8} = 5.225$

36. (C) Mean $= \frac{\sum x}{n} = \frac{1{,}800}{10} = 180$

37. (C) Mean $= \frac{\sum x}{n} = \frac{-280}{10} = -28$

38. (B) The median is $\frac{20+30}{2} = 25$. The median is the middle value or the arithmetic average of the two middle values in an *ordered* set of data. To find the median, first put the data values in order from least to greatest (or greatest to least): 15, 15, 15, 17, 20, 30, 30, 33, 45, 60. Next, locate the middle data value. If there is no single middle data value, compute the arithmetic average of the two middle data values: $\frac{20+30}{2} = 25$.

39. (D) Put the data in order: –4, –4, 11, 19, 25

Median = 11, the middle data value

40. (B) Put the data in order: 2.5, 4.7, 4.7, 4.9, 5.6, 5.6, 6.5, 7.3

$$\text{Median} = \frac{4.9 + 5.6}{2} = 5.250$$

41. (B) Put the data in order: 0, 0, 0, 0, 0, 100, 100, 100, 100, 1400

$$\text{Median} = \frac{0 + 100}{2} = 50$$

42. (B) Put the data in order: −60, −45, −33, −30, −30, −20, −17, −15, −15, −15

$$\text{Median} = \frac{(-30) + (-20)}{2} = -25$$

43. (A) The mode is 15 because it occurs with the greatest frequency. The mode is the data value (or values) that occurs with the greatest frequency in a data set. If only one data value occurs with the greatest frequency, then the data set is unimodal. If exactly two data values occur with the same frequency that is greater than any of the other frequencies, then the data set is bimodal. If more than two data values occur with the same frequency that is greater than any of the other frequencies, then the data set is multimodal. If each data value occurs the same number of times, the data set has no mode.

44. (D) For the data set given, each value occurs two times, so the data set has no mode.

45. (C) The data set has two modes: 4.7 and 5.6.

46. (A) 0 is the mode.

47. (A) −15 is the mode.

48. (A) The student's exam score average is

$$\frac{10\%(80) + 15\%(70) + 20\%(55) + 25\%(90) + 30\%(60)}{100\%} = 70$$

49. (A) The mean of the student's unit exams is $\frac{78 + 81 + 75}{3} = 78$. Therefore, the student's numerical course grade is

$$\frac{50\%(78) + 10\%(92) + 40\%(75)}{100\%} = 78.2$$

50. (D) Let x = the number. Then solving $\frac{5 + 8 + 10 + 12 + x}{5} = 10$ yields the following:

$$\begin{aligned} 5 + 8 + 10 + 12 + x &= 50 \\ 35 + x &= 50 \\ x &= 15 \end{aligned}$$

51. **(C)** Let x = the student's score on the fourth test. Then solving $\frac{77+91+94+x}{4}=90$ yields the following:

$$\begin{aligned} 77+91+94+x &= 360 \\ 262+x &= 360 \\ x &= 98 \end{aligned}$$

52. **(D)** According to the line graph, monthly revenues were \$140,000 in January, \$100,000 in February, \$120,000 in March, \$60,000 in April, \$140,000 in May, and \$120,000 in June, totaling \$680,000 for the period. Use the formula for a mean:

$$\frac{\sum x}{n}=\frac{\$680{,}000}{6}=\$113{,}333.33\approx\$113{,}000$$

53. **(B)** According to the bar graph shown, the new product received 8 ratings of 0, 4 ratings of 1, 8 ratings of 2, 6 ratings of 3, 19 ratings of 4, and 3 ratings of 5. In an ordered set of n data values, the location of the median is the $\frac{n+1}{2}$ position. For these data, $\frac{n+1}{2}=\frac{48+1}{2}=24.5$, so the median is halfway between the 24th and the 25th data value. From the information in the graph, you can determine that the 24th data value = 25th data value = 3, so the median is 3.

54. **(C)** The modal rating of the new product is 4, because this rating has the highest frequency.

55. **(D)** Mean starting weight in pounds $=\frac{\sum x}{n}=\frac{5{,}250}{36}=145.833\ldots\approx145.8$.

56. **(C)** For these data, $\frac{n+1}{2}=\frac{36+1}{2}=18.5$, so the median is halfway between the 18th and 19th data values. From the stem-and-leaf plot, you can determine that the 18th data value is 145 and the 19th data value is 146, so the median starting weight in pounds is $\frac{145+146}{2}=145.5$.

57. **(B)** The modal starting weight in pounds is 145 because it occurs with the greatest frequency.

58. **(B)** According to the dot plot shown, 3 customers waited 5 minutes, 2 customers waited 6 minutes, 1 customer waited 8 minutes, 4 customers waited 10 minutes, 2 customers waited 13 minutes, 3 customers waited 15 minutes, and 1 customer waited 19 minutes. Solve to find the mean number of minutes waited in line:

$$\frac{\sum x}{n}=\frac{165}{16}=10.3125\approx10.3$$

59. (A) For these data, $\frac{n+1}{2} = \frac{16+1}{2} = 8.5$, so the median is halfway between the 8th and 9th data values. From the dot plot, you can determine that the 8th data value = 9th data value = 10, so the median is 10.

60. (B) The modal number of minutes waited in line is 10 because it occurs with the greatest frequency.

61. (B) Using the results of questions 58 and 59, the mean lies slightly to the right of the median. Therefore, the distribution is slightly positively skewed.

62. (C) The mode is the most appropriate measure of central tendency because the data are nominal level.

63. (B) The median is the most appropriate measure of central tendency because the distribution is skewed.

64. (A) The mean is the most appropriate measure of central tendency because the distribution is symmetrical.

65. (A) Because the distribution is positively skewed, the mean is greater than the median. None of the other statements are true.

66. (B) Because the distribution is negatively skewed, the median is greater than the mean. None of the other statements are true.

67. (A) Only the mean (I.) is very sensitive to outliers.

68. (C) μ is the symbol commonly used for the population mean.

69. (D) $\bar{x}$ is the symbol commonly used for the sample mean.

70. (C) The distribution is symmetric because the data values are evenly distributed about the center, and the mean, median, and mode are equal to the same value (35).

Chapter 4: Measures of Variability

71. (D) Range = greatest value – least value = 50 – 0 = 50

72. (D) Range = 25 – (–4) = 29

73. (C) Range = –15 – (–60) = 45

74. (D) According to the line graph, the greatest revenues were $140,000 (in January and May) and the least revenues were $60,000 (in April). Thus, the range is $140,000 – $60,000 = $80,000.

75. (D) According to the stem-and-leaf plot, the greatest starting weight is 169, and the least is 123. Thus, the range is 169 – 123 = 46.

76. (A) According to the dot plot, the greatest number of minutes customers waited is 19, and the least is 5, so the range is 19 – 5 = 14.

77. (A) Both variance and standard deviation provide a measure of the variability about the mean.

78. (B) σ^2 is the symbol commonly used for the population variance.

79. (D) s^2 is the symbol commonly used for the sample variance.

80. (A) σ is the symbol commonly used for the population standard deviation.

81. (C) s is the symbol commonly used for the sample standard deviation.

82. (B) Mean $= \mu = \frac{\sum x}{N} = \frac{128}{5} = 25.6$

$$\text{Variance} = \sigma^2 = \frac{\sum (x_i - \mu)^2}{N}$$

$$= \frac{(15-25.6)^2 + (33-25.6)^2 + (30-25.6)^2 + (50-25.6)^2 + (0-25.6)^2}{5} \approx 287.440$$

83. (A) Mean $= \bar{x} = \frac{\sum x}{n} = \frac{128}{5} = 25.6$

$$\text{Variance} = s^2 = \frac{\sum (x_i - \bar{x})^2}{n-1}$$

$$= \frac{(15-25.6)^2 + (33-25.6)^2 + (30-25.6)^2 + (50-25.6)^2 + (0-25.6)^2}{4} \approx 359.300$$

84. (D) Mean $= \mu = \frac{\sum x}{N} = \frac{51}{6} = 8.5$

$$\text{Standard deviation} = \sigma = \sqrt{\frac{\sum (x_i - \mu)^2}{N}}$$

$$= \sqrt{\frac{(-4-8.5)^2 + (25-8.5)^2 + (-4-8.5)^2 + (11-8.5)^2 + (19-8.5)^2 + (4-8.5)^2}{6}}$$

≈ 10.97

85. **(C)** Mean $= \bar{x} = \dfrac{\sum x}{n} = \dfrac{51}{6} = 8.5$

Standard deviation $= s = \sqrt{\dfrac{\sum (x_i - \bar{x})^2}{n-1}}$

$$= \sqrt{\frac{(-4-8.5)^2 + (25-8.5)^2 + (-4-8.5)^2 + (11-8.5)^2 + (19-8.5)^2 + (4-8.5)^2}{5}}$$

≈ 12.01

86. **(C)** Mean $= \mu = \dfrac{\sum x}{N} = \dfrac{41.8}{8} = 5.225$

Variance $= \sigma^2 = \dfrac{\sum (x_i - \mu)^2}{N}$

$$= \frac{(4.7-5.225)^2 + (5.6-5.225)^2 + (2.5-5.225)^2 + (4.9-5.225)^2}{8}$$
$$\frac{+(7.3-5.225)^2 + (4.7-5.225)^2 + (5.6-5.225)^2 + (6.5-5.225)^2}{8} \approx 1.787$$

87. **(D)** Mean $= \bar{x} = \dfrac{\sum x}{n} = \dfrac{41.8}{8} = 5.225$

Variance $= s^2 = \dfrac{\sum (x_i - \bar{x})^2}{n-1}$

$$= \frac{(4.7-5.225)^2 + (5.6-5.225)^2 + (2.5-5.225)^2 + (4.9-5.225)^2}{7}$$
$$\frac{+(7.3-5.225)^2 + (4.7-5.225)^2 + (5.6-5.225)^2 + (6.5-5.225)^2}{7} \approx 2.042$$

88. **(A)** Mean $= \mu = \dfrac{\sum x}{N} = \dfrac{41.8}{8} = 5.225$

Standard deviation $= \sigma = \sqrt{\dfrac{\sum (x_i - \mu)^2}{N}}$

$$= \sqrt{\frac{(4.7-5.225)^2 + (5.6-5.225)^2 + (2.5-5.225)^2 + (4.9-5.225)^2}{8} \frac{+(7.3-5.225)^2 + (4.7-5.225)^2 + (5.6-5.225)^2 + (6.5-5.225)^2}{8}} \approx 1.337$$

89. (B) Mean $= \bar{x} = \dfrac{\sum x}{n} = \dfrac{41.8}{8} = 5.225$

Standard deviation $= s = \sqrt{\dfrac{\sum (x_i - \bar{x})^2}{n-1}}$

$$= \sqrt{\dfrac{(4.7-5.225)^2+(5.6-5.225)^2+(2.5-5.225)^2+(4.9-5.225)^2}{7} \atop \dfrac{+(7.3-5.225)^2+(4.7-5.225)^2+(5.6-5.225)^2+(6.5-5.225)^2}{7}} \approx 1.429$$

90. (D) Mean $= \mu = \dfrac{\sum x}{N} = \dfrac{-300}{6} = -50$

Standard deviation $= \sigma = \dfrac{\sum (x_i - \mu)^2}{N}$

$$= \sqrt{\dfrac{(-70-(-50))^2+(-60-(-50))^2+(-50-(-50))^2}{6} \atop \dfrac{+(-50-(-50))^2+(-40-50)^2+(-30-(-50))^2}{6}} \approx \sqrt{166.667} \approx 12.910$$

91. (C) Mean $= \bar{x} = \dfrac{\sum x}{n} = \dfrac{-300}{6} = -50$

Standard deviation $= s = \dfrac{\sum (x_i - \bar{x})^2}{n-1}$

$$= \sqrt{\dfrac{(-70-(-50))^2+(-60-(-50))^2+(-50-(-50))^2}{5} \atop \dfrac{+(-50-(-50))^2+(-40-50)^2+(-30-(-50))^2}{5}} = \sqrt{200} \approx 14.142$$

92. (B) The standard deviation for the data shown in the line graph is \$30,111. According to the line graph, revenues were \$140,000 in January, \$100,000 in February, \$120,000 in March, \$60,000 in April, \$140,000 in May, and \$120,000 in June. Using a calculator, enter these data into list **L1**:

Press STAT ENTER 140000 ENTER 100000 ENTER 120000 ENTER 60000 ENTER 140000 ENTER 120000 ENTER
Your screen will show the following:

```
L1      | L2     | L3     1
140000  | ------ | ------
100000  |        |
120000  |        |
60000   |        |
140000  |        |
120000  |        |
L1(7)=
```

Press STAT and use the right-arrow key to highlight **CALC**:

Press ENTER to paste item `1 : 1-Var Stats` to the home screen:

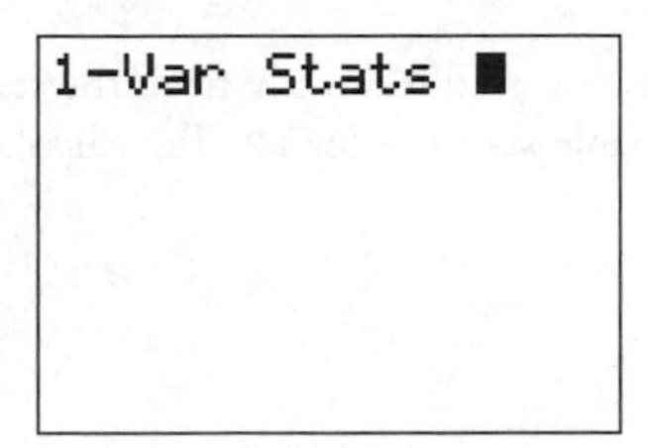

Go the List menu (2ND STAT), and press ENTER to select list **L1**.

Press ENTER to display the one-variable statistics for the data in list **L1**:

```
1-Var Stats
 x̄=113333.3333
 Σx=680000
 Σx²=8.16E10
 Sx=30110.90611
 σx=27487.37084
↓n=6
```

Sx is the sample standard deviation (and σx is the population standard deviation). Round 30110.90611 to $30,111, choice B.

Note: The ↓ on the last line of the screen indicates that you can use the down-arrow key to display more results. Use the down arrow to scroll down to show the rest of the statistics for the data in list **L1**:

```
1-Var Stats
↑n=6
 minX=60000
 Q1=100000
 Med=120000
 Q3=140000
 maxX=140000
```

93. (C) Proceed as in question 92. Enter the data (as you read it directly from the stem-and-leaf plot) into list **L2**, and then calculate one-variable statistics for **L2**. The calculator display will show the following:

```
1-Var Stats
 x̄=116.2
 Σx=1743
 Σx²=203469
 Sx=8.160882305
 σx=7.884161338
↓n=15
```

The sample standard deviation can be rounded to 8.16.

94. (B) Proceed as in question 92. Enter the data (as you read it directly from the dot plot) into list **L3**, and then calculate one-variable statistics for **L3**. The calculator display will show the following:

```
1-Var Stats
 x̄=10.3125
 Σx=165
 Σx²=1985
 Sx=4.346933785
 σx=4.208900539
↓n=16
```

Round the sample standard deviation to 4.3.

95. (A) The standard deviation is a good indicator of the reliability of the mean value. Group A's statistics have the least standard deviation. A smaller standard deviation means the data are more clustered around the mean. Being more clustered, the data set has fewer extreme values and thus has a more reliable mean.

96. (C) Each data value equals the mean (which is 50), so the variance is zero because there is no variability in the data set.

97. (C) Increasing each data value by 5 increases the mean by 5, but the standard deviation remains the same because the spread about the new mean is the same as the spread about the original mean.

98. (B) Eliminate choice D because the range is not a measure of location (central tendency). Eliminate choice C because, except for strictly nominal-level data, the mode is generally the least preferred measure of central tendency. Because the median is less influenced by outliers than is the mean, it is the best measure of location for describing skewed population data.

99. (A) Because the value of the mean is most stable from sample to sample (as compared with the mode and median), the mean is generally the best measure of location for describing sample data.

100. (C) Using a TI-84 Plus calculator, the one-variable statistics for the dot plot data are the following (see the answer to question 92 for the steps in the calculation):

```
1-Var Stats
 x̄=10.3125
 Σx=165
 Σx²=1985
 Sx=4.346933785
 σx=4.208900539
↓n=16
```

```
1-Var Stats
↑n=16
 minX=5
 Q1=6
 Med=10
 Q3=14
 maxX=19
```

Thus, the mean, $\bar{x}$, is 10.3125, the standard deviation, s, rounds to 4.347, and the median, Med, is 10. Substitute into the formula:

$$\text{Pearson's coefficient of skewness} = \frac{3(\bar{x} - \text{Med})}{s} = \frac{3(10.3125 - 10)}{4.347} = 0.216$$

101. (A) Using a TI-84 Plus calculator, the one-variable statistics for the data are the following:

```
1-Var Stats
 x̄=-180
 Σx=-1800
 Σx²=2000000
 Sx=431.5347289
 σx=409.3897898
↓n=10
```

```
1-Var Stats
↑n=10
 minX=-1400
 Q1=-100
 Med=-50
 Q3=0
 maxX=0
```

Thus, the mean, μ, is –180, the standard deviation, σ, rounds to 409.390, and the median, Med, is –50. Substitute into the formula:

$$\text{Pearson's coefficient of skewness} = \frac{3(\mu - \text{Med})}{\sigma} = \frac{3(-180 - (-50))}{409.390} = -0.953$$

102. (B) Use the formula for the coefficient of variation (CV):

$$CV = \frac{\sigma}{|\mu|} \times 100\% = \frac{4.5}{|360|} \times 100\% = \frac{4.5}{360} \times 100\% = 1.25\%$$

103. (B) Use the formula for the coefficient of variation (CV):

$$CV = \frac{s}{|\bar{x}|} \times 100\% = \frac{16.4}{|82|} \times 100\% = \frac{16.4}{82} \times 100\% = 20\%$$

104. (C) Compare the coefficients of variation of the four groups.

$$\text{Group A: } CV = \frac{s}{|\bar{x}|} \times 100\% = \frac{5}{25} \times 100\% = 20\%$$

$$\text{Group B: } CV = \frac{s}{|\bar{x}|} \times 100\% = \frac{7}{28} \times 100\% = 25\%$$

$$\text{Group C: } CV = \frac{s}{|\bar{x}|} \times 100\% = \frac{6}{31} \times 100\% \approx 19\%$$

$$\text{Group D: } CV = \frac{s}{|\bar{x}|} \times 100\% = \frac{9}{25} \times 100\% = 36\%$$

Relative to the mean, the data of Group C have the least relative variability.

Chapter 5: Percentiles and the Five-Number Summary

105. (C) 80th percentile $= \frac{33+45}{2} = 39$. The *R*th percentile is a value below which lie *R*% of the data values. To find the *R*th percentile for the data given, first order the data set: 15, 15, 15, 17, 20, 30, 30, 33, 45, 60. Next compute *k*, using the formula $k = n\left(\frac{R}{100}\right)$, where *n* is the number of data values:

$$k = n\left(\frac{R}{100}\right) = 10\left(\frac{80}{100}\right) = 8$$

If *k* is an integer, the *R*th percentile is the mean of the *k*th and (*k* + 1)th data values. If *k* is *not* an integer, round *k* up to the next greatest integer, say *l*, and then the *R*th percentile is the *l*th data value. Because *k* = 8 is an integer, find the mean of the 8th and 9th data values: $\frac{33+45}{2} = 39$.

106. (C) Order the data set: –4, –4, 4, 11, 19, 25

Compute *k*:

$$k = n\left(\frac{R}{100}\right) = 6\left(\frac{50}{100}\right) = 3$$

So the 50th percentile is halfway between the 3rd and 4th data values: $\frac{4+11}{2} = 7.5$.

107. (C) Order the data set: 2.5, 4.5, 4.5, 4.8, 5.6, 6.6, 6.6, 7.3

Compute *k*:

$$k = n\left(\frac{R}{100}\right) = 8\left(\frac{70}{100}\right) = 5.6$$

Because 5.6 is not an integer, round it up to *l* = 6. Thus, the 70th percentile is the 6th data value: 6.6.

108. (C) Order the data set: 0, 0, 0, 0, 0, 100, 100, 100, 100, 1400

Compute *k*:

$$k = n\left(\frac{R}{100}\right) = 10\left(\frac{90}{100}\right) = 9$$

The 90th percentile is halfway between the 9th and 10th data values: $\frac{100+1,400}{2} = 750$.

109. (D) Order the data set: −60, −45, −33, −30, −30, −20, −17, −15, −15, −15

Compute k:

$$k = n\left(\frac{R}{100}\right) = 10\left(\frac{35}{100}\right) = 3.5$$

Round up 3.5 to get $l = 4$. Thus, the 35th percentile is the 4th data value: −30.

110. (B)

$$k = n\left(\frac{R}{100}\right) = 48\left(\frac{25}{100}\right) = 12$$

The 25th percentile is halfway between the 12th and 13th data values: $\frac{1+2}{2} = 1.5$

111. (B)

Compute k:

$$k = n\left(\frac{R}{100}\right) = 36\left(\frac{95}{100}\right) = 34.2$$

Round up 34.2 to get $l = 35$. Thus, the 95th percentile is the 35th data value: 165.

112. (C)

$$k = n\left(\frac{R}{100}\right) = 16\left(\frac{37.5}{100}\right) = 6$$

The 37.5th percentile is halfway between the 6th and 7th data values: $\frac{8+10}{2} = 9$.

113. (C) Because the 60th percentile is $57,500, 60 % of the employees have salaries that are less than $57,500.

114. (B) Because 60% of the employees have salaries that are less than $57,500, 40% of 320 = 128 employees have salaries greater than $57,500.

115. (A) The median is the 50th percentile, so $48,000 has a percentile rank of 50.

116. (B) $57,500 is the 60th percentile, and $48,000 is the 50th percentile. Thus, 10% (60% – 50%) of 320 = 32 employees have salaries between $48,000 and $57,500.

117. (D) Additional information (such as the percentile rank of the mean of $55,000) is needed to determine the answer.

118. (A) The *z*-score (also called *standard score*) for a population data value is its distance in standard deviations from the population's mean:

$$z\text{-score} = \frac{\text{data value} - \text{mean}}{\text{standard deviation}} = \frac{x - \mu}{\sigma}.$$

The *z*-score expresses the position of a data value relative to the mean, which has *z*-score 0. Data values with negative *z*-scores lie to the left of the mean, and those with positive *z*-scores lie to the right of the mean.

Substituting into the formula, the *z*-score for $57,500 is

$$z\text{-score} = \frac{x - \mu}{\sigma} = \frac{\$57{,}500 - \$55{,}000}{\$4{,}000} = 0.625.$$

119. (A)

$$z\text{-score} = \frac{x - \mu}{\sigma} = \frac{\$48{,}000 - \$55{,}000}{\$4{,}000} = -1.75$$

120. (C) Substituting into the formula, $x = \mu + (z\text{-score})(\sigma) = \$55{,}000 + (1.25)(\$4{,}000) = \$60{,}000$

121. (B) Substituting into the formula (omitting units for convenience), $x = \mu + (z\text{-score})(\sigma) = 80 + (-1.5)(14) = 59$.

122. (D) Substituting into the formula (omitting units for convenience), $x = \mu + (z\text{-score})(\sigma) = 80 + (0.5)(14) = 87$.

123. (A) Compute the *z*-scores for each exam.

$$\text{Exam 1: } z\text{-score} = \frac{\text{exam score} - \text{mean}}{\text{standard deviation}} = \frac{75 - 65}{5} = 2$$

$$\text{Exam 2: } z\text{-score} = \frac{\text{exam score} - \text{mean}}{\text{standard deviation}} = \frac{87 - 88}{2} = -0.5$$

$$\text{Exam 3: } z\text{-score} = \frac{\text{exam score} - \text{mean}}{\text{standard deviation}} = \frac{92 - 86}{4} = 1.5$$

$$\text{Exam 4: } z\text{-score} = \frac{\text{exam score} - \text{mean}}{\text{standard deviation}} = \frac{70 - 60}{10} = 1$$

With a score that is 2 standard deviations above the mean, the student performed best on Exam 1 relative to the performance of the student's classmates.

124. (C) A box plot, also called box-and-whiskers plot, graphically displays the following (in this order): the minimum value (Min), the 25th percentile (Q_1, the lower quartile), the median (Med), the 75th percentile (Q_3, the upper quartile), and the maximum value (Max) of a data set. These values constitute the five-number summary for the data. In the box plot shown, Med = 10.5.

125. (A) Range = Max – Min = 17.3 – 6.1 = 11.2

126. (D) $\text{IQR} = Q_3 - Q_1 = 12.1 - 8.2 = 3.9$. The interquartile range (IQR) is the difference between the first and third quartiles: $\text{IQR} = Q_3 - Q_1$. It contains the center 50 percent of the data.

127. (A) Data values that fall outside the interval $(Q_1 - 1.5 \times \text{IQR},\ Q_3 + 1.5 \times \text{IQR}) = (8.2 - 1.5 \times 3.9,\ 12.1 + 1.5 \times 3.9) = (2.35,\ 17.95)$ are outliers. None of the data values for the box plot shown fall outside this interval, so there are no outliers in the data.

128. (B) The minimum value is approximately −13, so eliminate choices A and D. The maximum value is approximately 10.2, so eliminate choice C. Thus, choice B is the best match for the box plot.

129. (B) $\text{IQR} = Q_3 - Q_1 = 7.5 - (-10.2) = 17.7$

130. (B) Compute the summary statistics using a TI-84 Plus calculator.

```
1-Var Stats
↑n=48
 minX=0
 Q1=1.5
 Med=3
 Q3=4
 maxX=5
```

$\text{IQR} = Q_3 - Q_1 = 4 - 1.5 = 2.5$

131. (D) Compute the summary statistics using a TI-84 Plus calculator.

```
1-Var Stats
↑n=36
 minX=123
 Q1=136
 Med=145.5
 Q3=155.5
 maxX=169
```

$\text{IQR} = Q_3 - Q_1 = 155.5 - 136 = 19.5$

132. (B) Compute the summary statistics using a TI-84 Plus calculator.

```
1-Var Stats
↑n=16
 minX=5
 Q1=6
 Med=10
 Q3=14
 maxX=19
```

$\text{IQR} = Q_3 - Q_1 = 14 - 6 = 8$

133. (A) Data values that fall outside the interval $(Q_1 - 1.5 \times \text{IQR},\ Q_3 + 1.5 \times \text{IQR}) = (6 - 1.5 \times 8,\ 14 + 1.5 \times 8) = (-6,\ 26)$ are outliers. None of the data values for the dot plot shown fall outside this interval, so there are no outliers in the data.

134. (C) Outliers signal a concern that should be investigated. They should not be ignored.

Chapter 6: Counting Techniques

135. (D) The *fundamental counting principle* states that for a sequence of k tasks, if a first task can be done in any one of n_1 different ways, and after that task is completed, a second task can be done in any one of n_2 different ways, and after the first two tasks have been completed, a third task can be done in any one of n_3 different ways, and so on to the kth task, which can be done in any one of n_k different ways, then the total number of different ways the sequence of k tasks can be done is $n_1 \times n_2 \times n_3 \times \cdots \times n_k$. Therefore, the number of possible different meals is $8 \cdot 5 \cdot 7 = 280$.

136. (A) $10 \cdot 10 \cdot 10 \cdot 10 \cdot 10 = 100{,}000$ different possible codes

137. (B) $30 \cdot 29 \cdot 28 = 24{,}360$ different ways

138. (C) There are 26 ways to pick each of the two letters and 10 ways to pick each of the six digits, so there are $26^2 \cdot 10^6 = 26 \cdot 26 \cdot 10 \cdot 10 \cdot 10 \cdot 10 \cdot 10 \cdot 10 = 676{,}000{,}000$ different ID numbers.

139. (B) $26^3 \cdot 10^3 = 26 \cdot 26 \cdot 26 \cdot 10 \cdot 10 \cdot 10 = 17{,}576{,}000$ different car license plate alphanumeric codes

140. (A) $10^4 = 10 \cdot 10 \cdot 10 \cdot 10 = 10{,}000$ possible telephone numbers

141. (B) $15^3 = 15 \cdot 15 \cdot 15 = 3{,}375$ possible three-letter codes

142. (D) $2 \cdot 2 \cdot 2 = 8$

143. (B) $3 \cdot 2 \cdot 1 = 6$ different ways

144. (A) A *permutation* is an ordered arrangement of a set of distinctly different items. The number of permutation of n items taken n at a time is ${}_nP_n = n! = n(n-1)(n-2)\cdots(2)(1)$. Therefore, the number of different ways to arrange the five letters is ${}_5P_5 = 5 \cdot 4 \cdot 3 \cdot 2 \cdot 1 = 120$.

145. (D) The people are not assigned to particular seats, but are arranged relative to one another only. Therefore, for instance, the four arrangements shown below are the same because the people (P1, P2, P3, and P4) are in the same order clockwise.

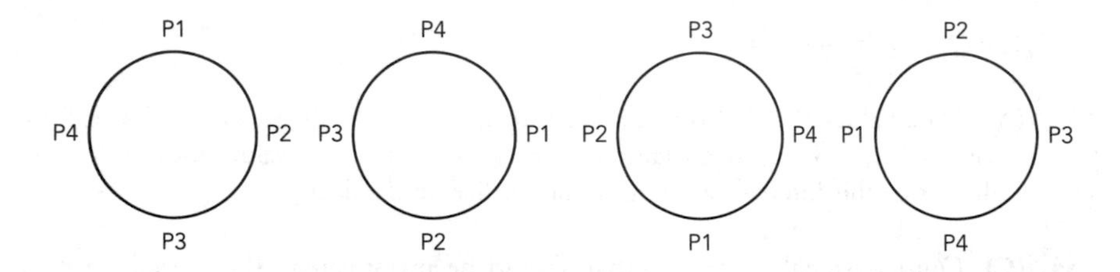

Hence, the position of P1 is immaterial. What counts is the positions of the other three people relative to P1. Therefore, keeping P1 fixed, there are ${}_3P_3 = 3 \cdot 2 \cdot 1 = 6$ different ways to arrange the other three people, so there are ${}_3P_3 = 6$ different arrangements of the four people in a circle.

146. (C) The permutation of n items taken r at a time is ${}_nP_r = \dfrac{n!}{(n-r)!} = n(n-1)(n-2)\cdots(n-r+1)$. Therefore, the number of different arrangements of the 10 people seated 8 at a time in 8 chairs is ${}_{10}P_8 = \dfrac{10!}{(10-8)!} = \dfrac{10!}{2!} = 10 \cdot 9 \cdot 8 \cdot 7 \cdot 6 \cdot 5 \cdot 4 \cdot 3 =$ 1,814,400.

147. (C) The following formula solves for the number of permutations of n items for which n_1 of the n items are alike, n_2 of the n items are alike, . . . , n_k of the n items are alike:

$$\frac{n!}{n_1!n_2!\cdots n_k!}$$

Therefore, there are $\dfrac{10!}{3!3!2!} = 50{,}400$ possible arrangements of the letters.

148. (C) $25 \cdot 24 \cdot 23 = 13{,}800$ different ways

149. (B) The permutation of n items taken r at a time is ${}_nP_r = \dfrac{n!}{(n-r)!} = n(n-1)(n-2)\cdots(n-r+1)$. Therefore, the number of different passwords is ${}_{10}P_6 = \dfrac{10!}{(10-6)!} = \dfrac{10!}{4!} = 10 \cdot 9 \cdot 8 \cdot 7 \cdot 6 \cdot 5 = 151{,}200$.

150. (B) A *combination* is an arrangement of a set of distinct objects in which different orderings of the same items are considered the same arrangement. The number of arrangements of r items selected from n distinct items is equal to ${}_nC_r = \frac{n!}{r!(n-r)!}$. Therefore, because the order of selection of the members of a committee is immaterial, the number of different ways the 3-member committee can be selected is ${}_{20}C_3 = \frac{20!}{3!(20-3)!} = \frac{20!}{3!17!} = 1{,}140$.

151. (C) Because the order in which the hands are dealt is immaterial, the number of possible 5-card hands is ${}_{52}C_5 = \frac{52!}{5!(52-5)!} = \frac{52!}{5!47!} = 2{,}598{,}960$.

152. (B)

$${}_{12}C_3 = \frac{12!}{3!(12-3)!} = \frac{12!}{3!9!} = 220 \text{ different ways}$$

153. (A)

$${}_{15}C_7 = \frac{15!}{7!(15-7)!} = \frac{15!}{7!8!} = 6{,}435 \text{ different ways}$$

154. (C) There are three tasks: Select 5 of the 8 shirts, select 4 of the 10 pairs of slacks, and select 3 of the 6 ties. Because the order of the selection of each of the clothing items is not important, use the combination formula to find the number of ways to make each selection. Following those calculations, use the fundamental counting principle to determine the number of ways to make the three selections: ${}_8C_5 \cdot {}_{10}C_4 \cdot {}_6C_3 = 235{,}200$ different ways.

Chapter 7: Basic Probability Concepts

155. (D) $52 \cdot 52 = 2{,}704$

156. (D) $2 \cdot 2 \cdot 2 \cdot 2 \cdot 2 \cdot 2 = 64$

157. (B) $P(\text{a number greater than 4 appears on the up face}) = \frac{1}{3}$. If all outcomes in the sample space S are equally likely (meaning each outcome is as likely to occur as any other outcome), the probability of an event E is given by the following formula:

$$P(E) = \frac{\text{number of outcomes favorable to } E}{\text{total number of outcomes in the sample space}}$$

On a six-sided die, you know that two faces are greater than 4, so you can solve for E = the event that a number greater than 4 appears on the up face in a single toss of a fair six-sided die:

$$P(E) = \frac{\text{number of outcomes favorable to } E}{\text{total number of outcomes in the sample space}} = \frac{2}{6} = \frac{1}{3}$$

158. (C) Using 1 for "1 appears on the up face," 2 for "2 appears on the up face," and so forth, $S = \{1, 2, 3, 4, 5, 6\}$. There are 5 favorable outcomes (2, 3, 4, 5, and 6) out of 6 total outcomes. Thus, $P(\text{prime or even}) = \frac{5}{6}$.

159. (A) $S = \{1, 2, 3, 4, 5, 6\}$, so there are no favorable outcomes in the sample space. Thus, $P(7) = \frac{0}{6} = 0$.

160. (C) Using H for "heads on the up face" and T for "tails on the up face," $S = \{\text{HH, HT, TH, TT}\}$. There are 3 favorable outcomes out of 4 total outcomes. Thus, $P(\text{at least one head}) = \frac{3}{4}$.

161. (B) Using ordered pairs to show the outcomes, where the first number indicates the number of dots on the first die, and the second number indicates the number of dots on the second die, there are six favorable outcomes: (1, 7), (7, 1), (2, 5), (5, 2), (3, 4), and (4, 3) out of $6 \cdot 6 = 36$ possible outcomes. Thus, $P(\text{sum is 7}) = \frac{6}{36} = \frac{1}{6}$.

162. (B) A standard deck of 52 playing cards consists of 4 suits: clubs, spades, hearts, and diamonds. Each suit has 13 cards marked 2, 3,..., 10, jack, queen, king, ace. Jacks, queens, and kings are face cards, so there are 12 favorable outcomes out of 52 possible outcomes. Thus, $P(\text{face card}) = \frac{12}{52} = \frac{3}{13}$.

163. (C) There are 16 favorable outcomes: 13 diamonds and 3 tens that are not diamonds. Thus, $P(\text{a diamond or a 10}) = \frac{16}{52} = \frac{4}{13}$.

164. (D) There are 17 favorable outcomes out of 23 total outcomes. Thus, $P(\text{black or red}) = \frac{17}{23}$.

165. (C) For any one question, there are 3 wrong answers out of 4 total answers, so the probability of guessing wrong on a particular question is $\frac{3}{4}$. Because the student is randomly guessing, each guess on a question is independent of the guesses on the other questions, so $P(\text{all five questions wrong}) = \frac{3}{4} \cdot \frac{3}{4} \cdot \frac{3}{4} \cdot \frac{3}{4} \cdot \frac{3}{4} = \frac{243}{1{,}024}$.

166. (C) $P(\text{defective}) = \frac{3}{30} = \frac{1}{10}$

167. (D) The complement of an event E, denoted $\overline{E}$, is the event that E does not occur: $P(\overline{E}) = 1 - P(E)$. In this problem, $P(\overline{E}) = 1 - P(E) = 1 - \frac{1}{3} = \frac{2}{3}$.

168. (C) $P(\bar{E}) = 1 - P(E) = 1 - \frac{5}{6} = \frac{1}{6}$

169. (C) $P(\bar{E}) = 1 - P(E) = 1 - \frac{17}{23} = \frac{6}{23}$

170. (A) $P(\bar{E}) = 1 - P(E) = 1 - \frac{3}{4} = \frac{1}{4}$

171. (B) $P(\bar{E}) = 1 - P(E) = 1 - \frac{1}{6} = \frac{5}{6}$

172. (B) $P(\text{not red}) = 1 - P(\text{red}) = 1 - \frac{10}{25} = \frac{15}{25} = \frac{3}{5}$

173. (C) $P(\text{at least one question correct}) = 1 - P(\text{none correct}) = 1 - \frac{243}{1{,}024}$

174. (D) $P(\text{not defective}) = 1 - P(\text{defective}) = 1 - \frac{1}{10} = \frac{9}{10}$

175. (B) Two events are mutually exclusive if they cannot occur at the same time; that is, they have no outcomes in common. Rolling a 3 and a number greater than 4 cannot occur at the same time on one toss of a fair six-sided die. The pair of events in choices A, C, and D can occur at the same time.

176. (D) $P(A \text{ or } B) = P(A) + P(B) - P(A \text{ and } B) = 0.5 + 0.3 - 0.06 = 0.74$

177. (B) $P(A \cup B) = P(A) + P(B) - P(A \cap B) = 0.4 + 0.1 - 0.05 = 0.45$

178. (C) $P(A \cup B) = P(A) + P(B) - P(A \cap B) = 0.65 + 0.22 - 0.08 = 0.79$

179. (B) $P(A \text{ or } B) = P(A) + P(B) - P(A \text{ and } B) = \frac{4}{52} + \frac{13}{52} - \frac{1}{52} = \frac{16}{52} = \frac{4}{13}$

180. (D) $P(\text{at least } A \text{ or } B) = P(A \cup B) = P(A) + P(B) - P(A \cap B) = \frac{3}{8} + \frac{5}{8} - 0 = 1$

181. (B) $P(\text{odd or less than 3}) = P(\text{odd}) + P(\text{less than 3}) - P(\text{odd and less than 3}) =$

$\frac{3}{6} + \frac{2}{6} - \frac{1}{6} = \frac{4}{6} = \frac{2}{3}$

182. (B) There are 4 aces, 12 spades that are not aces, and 12 diamonds that are not aces in the deck. Thus, $P(\text{ace or spade or diamond}) = \frac{4+12+12}{52} = \frac{28}{52} = \frac{7}{13}$.

183. (C) $S = \{\text{HHH, HHT, HTH, HTT, THH, THT, TTH, TTT}\}$, so $P(\text{at least two heads or number of heads is an odd number}) = \frac{7}{8}$.

184. (C) $S = \{\text{HHH, HHT, HTH, HTT, THH, THT, TTH, TTT}\}$, so $P(\text{all heads or all tails}) = \frac{2}{8} = \frac{1}{4}$.

185. (A) $P(A \text{ and } B) = P(A) + P(B) - P(A \text{ or } B) = 0.4 + 0.5 - 0.8 = 0.1$

186. (B) The probability of an event B, given that an event A has already occurred, is a conditional probability, denoted $P(B|A)$. $P(B|A) = \frac{P(A \cap B)}{P(A)}$. Substitute and evaluate:

$$P(B|A) = \frac{P(A \cap B)}{P(A)} = \frac{.32}{.35} = \frac{32}{35}$$

187. (A) Two events A and B are independent if the occurrence of one does not affect the probability of the occurrence of the other. When two events A and B are independent, $P(A) = P(A|B)$, $P(B) = P(B|A)$, and $P(A \cap B) = P(A)P(B)$. In choice A, because $P(X \cap Y) = \frac{1}{4}$ is the same as $P(X)\,P(Y) = \frac{1}{2} \cdot \frac{1}{2} = \frac{1}{4}$, the events X and Y are independent. Eliminate choice B because $P(X \cap Y) = 0.32$ is not the same as $P(X)\,P(Y) = (0.35)(0.63) = 0.2205$, so events X and Y are dependent. Eliminate choice C because $P(X \cap Y) = \frac{3}{22}$ is not the same as $P(X)\,P(Y) = \frac{1}{2} \cdot \frac{21}{55} = \frac{21}{110}$, so events X and Y are dependent. Eliminate choice D because $P(X) = 0.7$ does not equal $P(X|Y) = 0.9$, so events X and Y are dependent.

188. (A) The rolls are independent so the probability is $\frac{1}{2} \cdot \frac{2}{6} = \frac{1}{6}$

189. (D) Because the drawing is without replacement, the two events are dependent. Thus, the probability is $\frac{4}{52} \cdot \frac{12}{51} = \frac{4}{221}$

190. (A) The three flips are independent, so the probability is $\frac{1}{2} \cdot \frac{1}{2} \cdot \frac{1}{2} = \frac{1}{8}$

191. (B) Because the first card drawn is replaced before the second card is drawn, the two events are independent. Thus, the probability is $\frac{12}{52} \cdot \frac{8}{52} = \frac{6}{169}$

192. (B) Because the first marble drawn is replaced before the second marble is drawn, the two events are independent. Thus, the probability is $\frac{6}{14} = \frac{3}{7}$

193. (B) After 30 cell phones are removed, 70 remain of which 1 is defective. Thus, $P(\text{next cell phone selected is defective}) = \frac{1}{70}$.

194. (A) Fill in the missing probabilities.

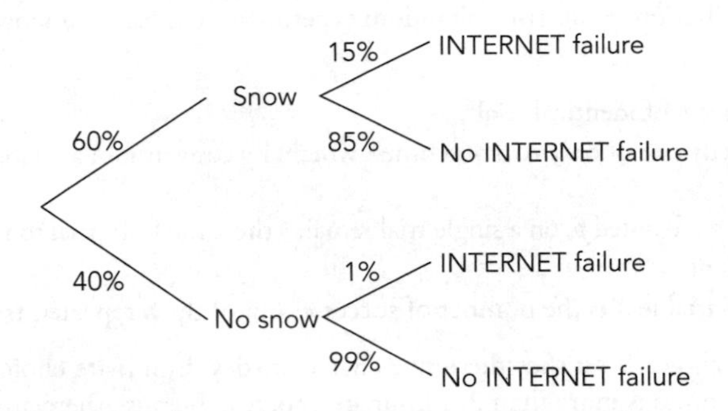

$P(\text{snow and Internet failure}) = P(\text{snow})P(\text{Internet failure}|\text{snow}) = (60\%)(15\%) = 9\%$

195. (B) Complete the table by filling in the row and column totals.

Residence Status of Senior Students ($n = 250$)

	On Campus	Off Campus	Total
Female	52	86	138
Male	38	74	112
Total	90	160	250

Method 1: You can work the problem using the formula for conditional probability:

$P(\text{resides on campus, given that the student selected is a male student})$

$$= P(\text{on campus}|\text{male}) = \frac{P(\text{on campus} \cap \text{male})}{P(\text{male})} = \frac{38/250}{112/250} = \frac{38}{112} \approx 0.34$$

Method 2: Because you know the student is male (it's given), another way to work this problem is to observe that 38 of the 112 male students reside on campus:

$$P(\text{on campus}|\text{male}) = \frac{38}{112} \approx 0.34$$

196. (A) Complete the table by filling in the row and column totals.

Residence Status of Senior Students ($n = 250$)

	On Campus	Off Campus	Total
Female	52	86	138
Male	38	74	112
Total	90	160	250

$$P(\text{off campus} \cap \text{female}) = \frac{86}{250} \approx 0.34$$

Chapter 8: Binomial Distribution

197. (B) A binomial distribution results from a random experiment that has the following characteristics:

- There are a fixed number n of identical trials.
- Each trial results in exactly one of only two outcomes, which, by convention, are labeled "success" and "failure."
- The probability of success, denoted p, on a single trial remains the same from trial to trial.
- The trials are independent.
- The binomial random variable X is the number of successes out of the n repeated trials.

Only the experiment in choice B satisfies these five characteristics. Eliminate choice A because the number of outcomes is more than 2. Eliminate choice C because there are no trials as such. Eliminate choice D because the probability of success is not constant from trial to trial seeing as the selections are made without replacement.

Note: Often, in real life, a binomial distribution can serve as a satisfactory model for the probability distribution of an experiment if, *for all practical purposes*, the experiment satisfies the properties of a binomial experiment. For these situations, you must use your own judgment when checking whether the five characteristics of a binomial distribution are satisfied.

198. (D) Only choice D fails to satisfy the characteristics of a binomial experiment. The number of outcomes is more than 2, and the drawings are done without replacement.

199. (C) $\mu = np = 8(0.2) = 1.6$

200. (B) $\sigma^2 = np(1-p) = 8(0.2)(0.8) = 1.28$

201. (A) $\sigma = \sqrt{np(1-p)} = \sqrt{8(0.2)(0.8)} \approx 1.13$

202. (D) $\mu = np = 15(0.9) = 13.5$

203. (B) $\sigma^2 = np(1-p) = 15(0.9)(0.1) = 1.35$

204. (A) $\sigma = \sqrt{np(1-p)} = \sqrt{15(0.9)(0.1)} \approx 1.16$

205. (B) $\mu = np = 100(0.5) = 50$

206. (C) $\sigma^2 = np(1-p) = 100(0.5)(0.5) = 25$

207. (D) $\sigma = \sqrt{np(1-p)} = \sqrt{100(0.5)(0.5)} = 5$

208. (C) $\mu = np = 300(0.25) = 75$

209. (B) $\sigma^2 = np(1-p) = 300(0.25)(0.75) = 56.25$

210. (A) $\sigma = \sqrt{np(1-p)} = \sqrt{300(0.25)(0.75)} \approx 7.5$

211. (A) $\mu = np = 20(0.15) = 3$

212. (A) The formula for the probability of observing exactly x successes in n trials of a binomial experiment is $P(X = x) = \binom{n}{x} p^x q^{n-x}$, where p is the probability of success on a single trial, q is $1 - p$, and $\binom{n}{x}$ is ${}_nC_x = \frac{n!}{x!(n-x)!}$. Thus, $P(X < 1) = P(X = 0) =$ $\frac{5!}{0!5!}(.2)^0(.8)^5 = 0.32768 \approx 0.328.$

213. (B) Refer to the answer to question 212 for the formula. $P(X \leq 1) = P(X = 0) +$ $P(X = 1) = \frac{5!}{0!5!}(.2)^0(.8)^5 + \frac{5!}{1!4!}(.2)^1(.8)^4 = 0.73728 \approx 0.737$

214. (C) Refer to the answer to question 212 for the formula. $P(X > 8) = P(X = 9) +$ $P(X = 10) = \frac{10!}{9!1!}(.5)^9(.5)^1 + \frac{10!}{10!0!}(.5)^{10}(.5)^0 = 0.0107421875 \approx 0.011$

215. (B) $P(X = 2) = 0.441$. To find this solution with the TI-84 Plus, press 2ND VARS to access the **DISTR** menu:

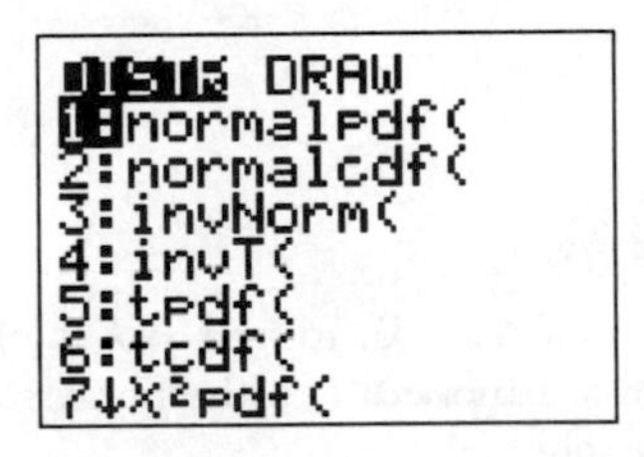

Using the down-arrow key, scroll down until you see items A and B:

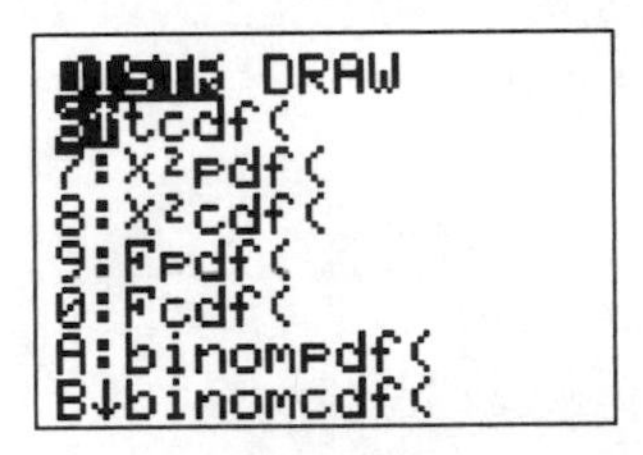

To find $P(X = x)$, the probability of exactly x successes in n trials, use item **A : binompdf(**. Scroll to item A. Press ENTER to paste **binompdf(** to the home screen. The open left parenthesis prompts you to enter values for n, p, and x, where n is the number of trials, p is the probability of a success on a single trial, and x is the number of successes in the n trials. For $P(X = 2)$, type **3,0.7,2)**. Press ENTER. The display shows the solution:

```
binompdf(3,0.7,2
)
                .441
```

216. (D) $P(X \le 2) = 0.657$. To find this solution with the TI-84 Plus, press 2ND VARS to access the **DISTR** menu. To find $P(X \le x)$, the cumulative probabilities from 0 up to and including x successes in n trials, use item **B : binomcdf(**. Scroll to item B. Press ENTER to paste **binomcdf(** to the home screen. The open left parenthesis prompts you to enter values for n, p, and x. For $P(X \le 2)$, type **3,0.7,2)**. Press ENTER. The display shows the solution:

```
binomcdf(3,0.7,2
)
                .657
```

217. (A) $P(X < 2) = 0.216$. Because $P(X < x) = P(X \le x - 1)$, to find $P(X < x)$, use **binomcdf(n,p,x-1)**. For $P(X < 2) = P(X \le 1)$, paste **binomcdf(** to the home screen. Type **3,0.7,1)**. Press ENTER. The display shows the solution:

```
binomcdf(3,0.7,1
)
                .216
```

218. **(A)** $P(X > 2) = 0.343$. Because $P(X > x) = 1 - P(X \leq x)$, to find $P(X > x)$, use `1-binomcdf(n, p, x)`. For $P(X > 2) = 1 - P(X \leq 2)$, type `1-` and then paste `binomcdf(` to the home screen. Next, type `3,0.7,2)`. Press ENTER. The display shows the solution:

```
1-binomcdf(3,0.7
,2)
              .343
```

219. **(D)** $P(X \geq 2) = 0.784$. Because $P(X \geq x) = 1 - P(X \leq x - 1)$, to find $P(X \geq x)$, use `1-binomcdf(n, p, x-1)`. For $P(X \geq 2) = 1 - P(X \leq 1)$, type `1-` and then paste `binomcdf(` to the home screen. Next, type `3,0.7,1)`. Press ENTER. The display shows the solution:

```
1-binomcdf(3,0.7
,1)
              .784
```

220. **(C)** $P(X = 5) \approx 0.037$

```
binompdf(6,0.4,5
)
           .036864
```

221. **(D)** $P(X \geq 1) \approx 0.999$

```
1-binomcdf(10,0.
5,0)
       .9990234375
```

222. (A) $P(X$ is less than $1) = P(X < 1) = P(X \leq 0) \approx 0.035$

```
binomcdf(15,0.2,
0)
         .0351843721
```

223. (D) $P(X$ not less than $8) = P(X \geq 8) = 1 - P(X \leq 7) \approx 0.930$

```
1-binomcdf(10,0.
9,7)
         .9298091736
```

224. (C) $P(X$ exceeds $8) = P(X > 8) = 1 - P(X \leq 8) \approx 0.011$

```
1-binomcdf(10,0.
5,8)
         .0107421875
```

225. (B) $P(X$ is less than or equal to $3) = P(X \leq 3) \approx 0.795$

```
binomcdf(12,0.2,
3)
         .7945689495
```

226. (A) $P(X$ is exactly $17) = 0.154$. The binomial probabilities table in Appendix A gives binomial probabilities (rounded to three decimal places) for $P(X = x)$ for various values of n, p, and x. To use the table, first locate n in the leftmost column. Then, for that n, read down to find the row in which the corresponding value of x appears. Now read across that row to the desired value for p. Thus, when $n = 19$ and $p = 0.8$, $P(X$ is exactly $17) = p(X = 17) = 0.154$.

227. (A) When $n = 11$ and $p = 0.4$, $P(X$ is at most $2) = p(X \leq 2) = p(X = 0) + p(X = 1) + p(X = 2) = 0.004 + 0.027 + 0.089 = 0.120$

228. (D) When $n = 9$ and $p = 0.7$, $P(X$ is no less than $6) = p(X \geq 6) = p(X = 6) + p(X = 7) + p(X = 8) + p(X = 9) = 0.267 + 0.267 + 0.156 + 0.040 = 0.730$

229. (C) When $n = 20$ and $p = 0.9$, $P(X$ is greater than $18) = p(X > 18) = p(X = 19) + p(X = 20) = 0.270 + 0.122 = 0.392$

230. (C) When $n = 15$ and $p = 0.2$, $P(X$ is no more than $1) = P(X \leq 1) = P(X = 0) + P(X = 1) = 0.035 + 0.132 = 0.167$

231. (A) When $n = 16$ and $p = 0.7$, $P(X$ is at least $15) = p(X \geq 15) = p(X = 15) + p(X = 16) = 0.023 + 0.003 = 0.026$

232. (B) $n = 5, p = \frac{3}{6} = 0.5, P(X = 4) \approx 0.156$

```
binompdf(5,0.5,4
)
              .15625
```

233. (D) $n = 3, p = 0.5, P(X \leq 2) = 0.875$

```
binomcdf(3,0.5,2
)
                .875
```

234. (C) $n = 10, p = 0.4, P(X = 8) \approx 0.011$

```
binompdf(10,0.4,
8)
         .010616832
```

235. (C) $n = 5, p = 0.1, P(X = 2) \approx 0.073$

```
binompdf(5,0.1,2
)
                .0729
```

236. (A) $n = 5, p = \frac{1}{4} = 0.25, P(X = 5) = 0.0009765625 \approx 0.001$

```
binompdf(5,0.25,
5)
        9.765625E-4
```

237. (B) $n = 100, p = \frac{1}{6}, P(X \geq 20) \approx 0.220$

```
1-binomcdf(100,1
/6,19)
        .2197498382
```

238. (A) $n = 10, p = \frac{1}{5} = 0.2, P(X \geq 3) \approx 0.322$

```
1-binomcdf(10,0.
2,2)
         .322200474
```

Chapter 9: Normal Distribution

239. (C) The **normal distribution** or **normal curve** is the name given to a family of distributions that have the following properties in common.

1. The equation for the graph is given by $y = \frac{e^{-\frac{1}{2}\left(\frac{x-\mu}{\sigma}\right)^2}}{\sigma\sqrt{2\pi}}$.
2. The distribution is completely defined by two parameters: its mean, μ, and standard deviation, σ.
3. The distribution is bell-shaped and symmetric about its mean, μ.
4. The mean, median, and mode coincide.
5. The distribution is continuous extending from $-\infty$ to ∞.
6. The standard deviation, σ, is the distance from μ to where the curvature of the graph changes on either side.
7. The x-axis is a horizontal asymptote for the curve.
8. The total area bounded by the curve and the horizontal axis is 1, with 0.5 area to the left of μ and 0.5 area to the right of μ.
9. Because the total area bounded by the curve and the horizontal axis is 1, the area over a particular interval along the horizontal axis equals the probability that the normal random variable will assume a value in that interval.
10. About 68.26% of the values of a random variable that is normally distributed fall within 1 standard deviation of the mean, 95.44% fall within two standard deviations of the mean, and 99.74% fall within three standard deviations of the mean (68.26-95.44-99.74 property).

Thus, $P(\mu - 1\sigma < X < \mu + 1\sigma) = 0.6826$ (according to the 68.26-95.44-99.74 property).

240. (B) Because the total area under the normal curve is 1, the probability that X assumes a value more than 1 standard deviation from its mean = 1 – 0.6826 = 0.3174.

241. (A) $P(260 < X < 340) = P(300 - 2 \cdot 20 < X < 300 + 2 \cdot 20) = P(\mu - 2\sigma < X < \mu + 2\sigma) = 0.9544$ (according to the 68.26-95.44-99.74 property)

242. (B) $1 - P(260 < X < 340) = 1 - 0.9544$ (from question 241) $= 0.0456$

243. (D) $P(125 < X < 275) = P(200 - 3 \cdot 25 < X < 200 + 3 \cdot 25) = P(\mu - 3\sigma < X < \mu + 3\sigma) = 0.9974$ (according to the 68.26-95.44-99.74 property)

244. (C) Only the statement in choice C is always true because the probability that a normal random variable assumes any particular exact value is zero. The statements in the other answer choices are false.

245. (B) The *standard normal distribution*, denoted Z, is the special normal distribution that has mean $\mu = 0$ and standard deviation $\sigma = 1$. Due to its unique parameters, a point z along the horizontal axis of a standard normal distribution expresses the position of the value *relative to the mean*, with negative values lying to the left of the mean and positive values lying to the right of the mean. Furthermore, the z-value is in terms of standard deviations. For instance, the value –1 is one standard deviation below the mean and the value 1 is one standard deviation above the mean. Thus, $P(-1 < Z < 1) = 0.6826$ (according to the 68.26-95.44-99.74 property).

246. (C) $P(-2 < Z < 2) = 0.9544$ (according to the 68.26-95.44-99.74 property)

247. (D) $P(-3 < Z < 3) = 0.9974$ (according to the 68.26-95.44-99.74 property)

248. (C) The total area to the right of the mean of a normal curve is 0.5. Thus, $P(Z \geq 0) = 0.5$.

249. (A) Because the probability that a normal random variable assumes any particular exact value is zero, $P(Z = 0) = 0$.

250. (A) $P(Z < 1.65) = 0.9505$. The standard normal table in Appendix B gives the cumulative area bounded by the normal curve and the horizontal axis all the way from negative infinity ($-\infty$) up to a vertical cutoff line at a specific z-value. In other words, the entry in the table corresponding to a particular z-value is the area under the curve all the way to the left of that z-value. To find the entry for a particular z-value, say $z = 1.65$, write the z-value as a number that has exactly two decimal places. Next, split the z-value into two parts: the first part is the number to the tenths place, and the second part is the hundredths part of the number.

For $z = 1.65$, the first part is 1.6, and the hundredths part is 0.05. Look in the table for the intersection of the row labeled 1.6 and the column labeled 0.05. According to the table, $P(Z < 1.65) = 0.9505$.

251. (A) $P(Z < 0.08) = 0.5319$

252. (C) $P(Z \leq -1.99) = 0.0233$

253. (D) $P(Z \geq -1.04) = 1 - P(Z < -1.04) = 1 - 0.1492 = 0.8508$

254. (A) $P(Z > 1.85) = 1 - P(Z \leq 1.85) = 1 - 0.9678 = 0.0322$

255. (B) $P(-1.2 < Z < 2.4) = P(-1.20 < Z < 2.40) = P(Z < 2.40) - P(Z < -1.20) = 0.9918 - 0.1151 = 0.8767$

256. (C) $P(Z < -1.96) \approx 0.0250$. Press 2ND VARS to access the `DISTR` menu. Press **2** to paste **`normalcdf(`** to the home screen. This command uses the following syntax: **`normalcdf(`** (lowerbound, upperbound, μ, σ). For $P(Z < -1.96)$. Type **`-1E99,-1.96,0,1)`**. Press ENTER. The display shows the probability:

```
normalcdf(-1E99,
-1.96,0,1)
        .0249978252
█
```

Note: Use **`-1E99`** to approximate $-\infty$. The **`E`** key, labeled **`EE`**, is above the comma key on the TI-84 Plus and is used for scientific notation.

257. **(D)** $P(Z \geq -1.96) \approx 0.9750$

```
normalcdf(-1.96,
1E99)
      .9750021748
■
```

Notice that when finding probabilities for the standard normal curve, you can use the simplified syntax: **normalcdf(** (lowerbound, upperbound). *Caution:* Use this simplified form only when $\mu = 0$ and $\sigma = 1$.

258. **(B)** $P(-1.96 < Z < 0) \approx 0.4750$

```
normalcdf(-1.96,
0)
      .4750021754
```

259. **(B)** $P(0 < Z < 1.96) \approx 0.4750$

```
normalcdf(0,1.96
)
      .4750021743
■
```

260. **(C)** $P(-1.96 < Z < 1.96) \approx 0.9500$

```
normalcdf(-1.96,
1.96)
      .9500043497
```

261. **(A)** $P(Z < -3) \approx 0.0013$

```
normalcdf(-1E99,
-3)
      .0013499672
```

262. **(D)** $P(Z < 3) \approx 0.9987$

```
normalcdf(-1E99,
3)
      .9986500328
```

263. **(C)** $P(Z \leq 0) = 0.5000$

```
normalcdf(-1E99,
0)
      .5000000005
```

264. **(C)** $P(Z > 2.33) \approx 0.0099$

```
normalcdf(2.33,1
E99)
       .009903053
```

265. (A) $P(1.23 < Z \leq 1.87) \approx 0.0786$

```
normalcdf(1.23,1
.87)
        .0786067718
```

266. (A) $P(0 < Z < 1.65) \approx 0.4505$

```
normalcdf(0,1.65
)
        .4505285486
```

267. (C) $P(-1 < Z < 1) \approx 0.6827$

```
normalcdf(-1,1)
        .6826894809
```

268. (B) $P(-1.95 < Z < -1.28) \approx 0.0747$

```
normalcdf(-1.95,
-1.28)
        .0746846437
```

269. (C) Recall from question 105 that the *R*th percentile is a value below which lie *R*% of the data values. To determine the percentile designation of a *z*-score, find *R*%, the percent of the standard normal curve to the left of the *z*-score. The given *z*-score is the *R*th percentile of the standard normal distribution. Because $P(Z < -2.5) = 0.0062 = 0.62\%$, $z = -2.5$ is the 0.62th percentile.

270. (D) Because $P(Z < 2.5) = 0.9938 = 99.38\%$, $z = 2.5$ is the 99.38th percentile of the standard normal distribution.

271. (A) Because $P(Z < -1.49) = 0.0681 = 6.81\%$, $z = -1.49$ is the 6.81th percentile of the standard normal distribution.

272. (B) Because $P(Z < 1.64) = 0.9495 = 94.95\%$, $z = 1.64$ is the 94.95th percentile of the standard normal distribution.

273. (D) Because $P(Z < 0) = 0.5000 = 50\%$, $z = 0$ is the 50th percentile of the standard normal distribution.

274. (D) The 90th percentile of the standard normal distribution is $z \approx 1.28$. Press 2ND VARS to access the **DISTR** menu. Press 3 to paste **invNorm(** to the home screen. This command uses the following syntax:
invNorm((area to left in decimal form, μ,σ). For the 90th percentile, type **.90,0,1)**.
Press ENTER. The display shows the solution:

```
invNorm(.90,0,1)
          1.281551567
```

275. (A) The 20th percentile of the standard normal distribution is $z \approx -0.84$.

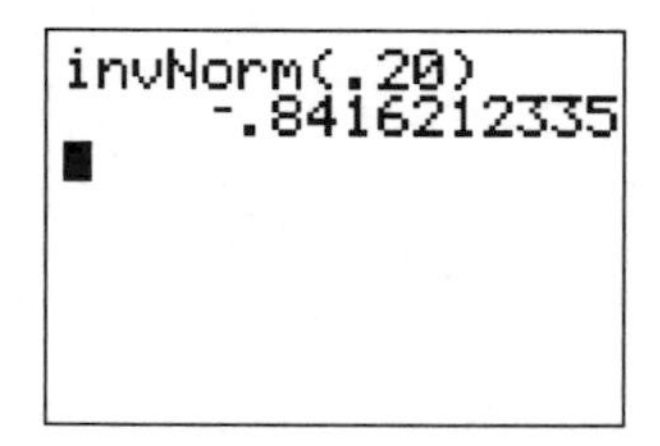

Notice that when finding percentiles of the standard normal distribution, you can use the simplified syntax **invNorm(** (area to left in decimal form). *Caution:* Use this simplified form only when $\mu = 0$ and $\sigma = 1$.

276. (B) Proceed as in the answer to question 274. The 14th percentile of the standard normal distribution is $z \approx -1.08$.

277. (C) Proceed as in the answer to question 274. The 60th percentile of the standard normal distribution is $z \approx 0.25$.

278. (B) Proceed as in the answer to question 274. The 3rd percentile of the standard normal distribution is $z \approx -1.88$.

279. (B) Proceed as in the answer to question 274. $z_0 \approx 1.16$

280. (A) Proceed as in the answer to question 274. $z_0 \approx -2.24$

281. (D) Proceed as in the answer to question 274. $z_0 \approx 2.49$

282. (A) The area to the right of z_0 is 0.975, so the area to its left is $1 - 0.975 = 0.025$. To find z_0 such that $P(Z < z_0) = 0.025$, proceed as in the answer to question 274. $z_0 = -1.96$

283. (C) The area to the right of z_0 is 0.119, so the area to its left is $1 - 0.119 = 0.881$. To find z_0 such that $P(Z < z_0) = 0.881$, proceed as in the answer to question 274. $z_0 = 1.18$

284. (A) Because the probability that a normal random variable assumes any particular exact value is zero, $P(X = 60) = 0$.

285. (C) The *z*-score is used to transform any normal random variable with mean μ and standard deviation σ into a standard normal random variable. The *z*-score, denoted *z*, for a given *x*-value is given by the formula

$$z = \frac{x - \mu}{\sigma}.$$

The transformation formula allows you to find probabilities for any normally distributed random variable. You simply convert *x*-values to *z*-scores, and then work the problem as a standard normal distribution problem.

Therefore, $P(X \leq 60) = P\left(Z \leq \frac{60 - 60}{10}\right) = P(Z \leq 0) = 0.5000.$

286. (B) $P(50 \leq X \leq 70) = P\left(\frac{50 - 60}{10} \leq Z \leq \frac{70 - 60}{10}\right) = P(-1 \leq Z \leq 1) = 0.6826$

287. (A) $P(X \leq 20) = P\left(Z \leq \frac{20 - 25}{5}\right) = P(Z \leq -1) = 0.1587$

288. (D) $P(150 \leq X \leq 250) = P\left(\frac{150 - 300}{50} \leq Z \leq \frac{250 - 300}{50}\right) = P(-3 \leq Z \leq -1) =$ 0.1573

289. (C) $P(X \geq 40.5) = P\left(Z \geq \frac{40.5 - 30}{6}\right) = P(Z \geq 1.75) = 1 - 0.9599 = 0.0401$

290. (D) $P(X \geq 100) = P\left(Z \geq \frac{100-120}{25}\right) = P(Z \geq -0.80) = 1 - 0.2119 = 0.7881$

291. (A) $P(X \leq 1900) = P\left(Z \leq \frac{1900-1600}{200}\right) = P(Z \leq 1.5) = 0.9332$

292. (B) $P(X \geq 1900) = P\left(Z \geq \frac{1900-1600}{200}\right) = P(Z \geq 1.5) = 1 - 0.9332 = 0.0668$

293. (C) $P(X > 130) = P\left(Z > \frac{130-100}{15}\right) = P(Z > 2) = 1 - 0.9772 = 0.0228$

294. (A) $P(X \geq 750) = P\left(Z \geq \frac{750-500}{100}\right) = P(Z \geq 2.5) = 1 - 0.9938 = 0.0062$

295. (B) $P(X \leq 19) = P\left(Z \leq \frac{19-16}{1.5}\right) = P(Z \leq 2) = 0.9772$

296. (C) $P(X \leq 189) = P\left(Z \leq \frac{189-160}{25}\right) = P(Z \leq 1.16) = 0.8770$

297. (D) $P(X \geq 88) = P\left(Z \geq \frac{88-100}{8}\right) = P(Z \geq -1.5) = 1 - 0.0668 = 0.9332$

298. (A) Refer to the answer to question 256 for the calculator steps. $P(106.6 < X < 129.4) \approx 0.7699 = 76.99\%$

```
normalcdf(106.6,
129.4,118,9.5)
        .7698605368
■
```

299. (B) Refer to the answer to question 256 for the calculator steps. $P(X \leq 6) \approx 0.9772$

```
normalcdf(-1E99,
6,5,0.5)
         .977249938
■
```

300. **(C)** Refer to the answer to question 256 for the calculator steps. $P(X > 9.5) \approx 0.0668$

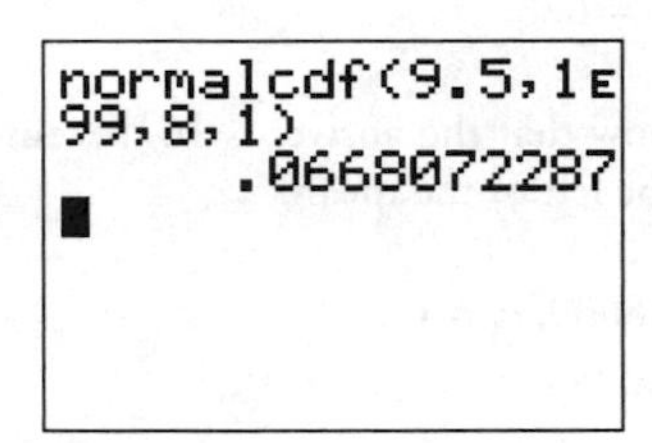

301. **(C)** Refer to the answer to question 256 for the calculator steps. $P(X \geq 500) \approx 0.3085 = 30.85\%$

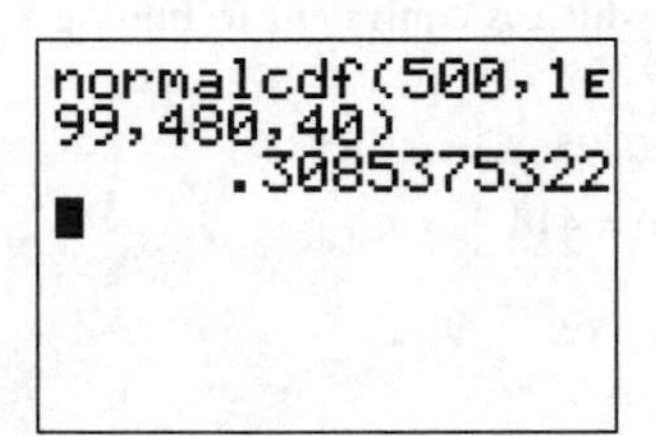

302. **(A)** Step 1. Find z_0 such that $P(Z \leq z_0) = 90\% = 0.9000$; $z_0 = 1.28$
Step 2. $x_0 = \mu + z_0\sigma = 100 + (1.28)(15) = 119.2$ is the 90th percentile.

303. **(B)** Step 1. Find z_0 such that $P(Z \leq z_0) = 20\% = 0.2000$; $z_0 = -0.84$
Step 2. $x_0 = \mu + z_0\sigma = 100 + (-0.84)(15) = 87.4$ is the 20th percentile.

304. **(A)** Because 50% of the data values lie to the left of the mean, 100 is the 50th percentile.

305. **(D)** Step 1. Find z_0 such that $P(Z \leq z_0) = 0.8771$; $z_0 = 1.16$
Step 2. $x_0 = \mu + z_0\sigma = 500 + (1.16)(100) = 616$

306. **(A)** Step 1. Find z_0 such that $P(Z \leq z_0) = 0.0124$; $z_0 = -2.24$
Step 2. $x_0 = \mu + z_0\sigma = 500 + (-2.24)(100) = 276$

307. **(D)** Step 1. Find z_0 such that $P(Z \leq z_0) = 0.9936$; $z_0 = 2.49$
Step 2. $x_0 = \mu + z_0\sigma = 500 + (2.49)(100) = 749$

308. **(C)** Step 1. Find z_0 such that $P(Z \leq z_0) = 0.025$; $z_0 = -1.96$
Step 2. $x_0 = \mu + z_0\sigma = 500 + (-1.96)(100) = 304$

309. **(D)** Step 1. Find z_0 such that $P(Z \leq z_0) = 0.881$; $z_0 = 1.18$
Step 2. $x_0 = \mu + z_0\sigma = 500 + (1.18)(100) = 618$

310. (D) Step 1. Find z_0 such that $P(Z \le z_0) = 0.9000$; $z_0 = 1.28$

Step 2. $x_0 = \mu + z_0\sigma = 500 + (1.28)(100) = 628$

311. (B) Without doing any calculations, you know that the answer is 16, because in a normal distribution, 50% of the data values lie to the left of the mean.

312. (A) Step 1. Find z_0 such that $P(Z \le z_0) = 0.8000$; $z_0 = 0.84$

Step 2. $x_0 = \mu + z_0\sigma = 160 + (0.84)(25) = 181$

313. (A) Step 1. Find z_0 such that $P(Z \le z_0) = 0.2500$; $z_0 = -0.67$

Step 2. $x_0 = \mu + z_0\sigma = 100 + (-0.67)(8) = 94.64 \approx 95$

314. (D) Find z_0 such that $P(Z \ge z_0) = 0.9505$, which is equivalent to finding z_0 such that $P(Z \le z_0) = 1 - 0.9505 = 0.0495$.

Step 1. Find z_0 such that $P(Z \le z_0) = 0.0495$; $z_0 = -1.65$

Step 2. $x_0 = \mu + z_0\sigma = 480 + (-1.65)(40) = 414$

Chapter 10: Estimation

315. (A) Because $n = 100 \ge 30$, the sampling distribution of $\overline{X}$ is approximately normal with mean $= \mu = 300$ and standard deviation $= \frac{\sigma}{\sqrt{n}} = \frac{50}{\sqrt{100}} = 5$. When you sample from a population, the numerical value of the mean of the sample is a random variable $\overline{X}$ that estimates the population mean. Like all random variables, $\overline{X}$ has a probability distribution. This distribution is called the *sampling distribution* of $\overline{X}$. The central limit theorem states that when sampling repeatedly from a population with mean μ and finite standard deviation σ using a fixed sample size n, the sampling distribution of $\overline{X}$ will approach a normal distribution with mean μ and standard deviation $\frac{\sigma}{\sqrt{n}}$ as the sample size n becomes large ($n \ge 30$).

Some important points to keep in mind about the central limit theorem are the following:

1. Two distributions are involved in the statement of the central limit theorem: (1) the distribution of the population that is being sampled and (2) the sampling distribution of $\overline{X}$.
2. Usually you do not know whether the distribution of the population is or is not normally distributed.
3. The central limit theorem is noteworthy because it states that if the sample size is large enough ($n > 30$), the sampling distribution of $\overline{X}$ is approximately normal *regardless* of the distribution of the population.
4. The mean of all the sample means is μ, the population mean.
5. The standard deviation of all the sample means is $\frac{\sigma}{\sqrt{n}}$, the population standard deviation divided by the square root of the sample size.
6. The sampling distribution of $\overline{X}$ is a theoretical concept, meaning that in practice *only one sample is taken and one sample mean is calculated.*

316. (B) Because $n = 36 \geq 30$, the sampling distribution of $\bar{X}$ is approximately normal with mean $= \mu = 60$ and standard deviation $= \frac{\sigma}{\sqrt{n}} = \frac{15}{\sqrt{36}} = 2.5$.

317. (C) When you sample from a normally distributed population, you do not need the central limit theorem because, regardless of the sample size, the sampling distribution of the sample mean is a normal distribution. For this question, because the sampling is from a normally distributed population, the sampling distribution of $\bar{X}$ has a normal distribution with mean $\mu = 25$ and standard deviation $= \frac{\sigma}{\sqrt{n}} = \frac{6}{\sqrt{9}} = 2$.

318. (D) The methods from Chapter 9 for dealing with normal distributions can be applied to problems involving the sampling distribution of $\bar{X}$. Just remember that when you are working with a mean from a sample, you use $\frac{\sigma}{\sqrt{n}}$ for the standard deviation of the normal distribution instead of σ.

For this question, according to the central limit theorem, the distribution of $\bar{X}$ is approximately normal with mean $\mu = 300$ and standard deviation $\frac{\sigma}{\sqrt{n}} = \frac{50}{\sqrt{100}} = 5$.

$$\text{Thus, } P(299 < \bar{X} < 301) = P\left(\frac{299-300}{50/\sqrt{100}} < Z < \frac{301-300}{50/\sqrt{100}}\right)$$

$$= P\left(\frac{299-300}{5} < Z < \frac{301-300}{5}\right) = P(-0.2 < Z < 0.2) = 0.1585$$

319. (A) $P(\bar{X} \geq 58) = P\left(Z \geq \frac{58-60}{15/\sqrt{36}}\right) = P\left(Z \geq \frac{58-60}{2.5}\right) = P(Z \geq -0.8) = 0.7881$

320. (B)

$$P(\bar{X} > 26.7) = P\left(Z > \frac{26.7-25}{6/\sqrt{9}}\right) = P\left(Z > \frac{26.7-25}{2}\right) = P(Z > 0.85) = 1 - 0.8023$$

$$= 0.1977$$

321. (A) $P(\bar{X} \leq 26.08) = P\left(Z \leq \frac{26.08-30}{12/\sqrt{36}}\right) = P\left(Z \leq \frac{26.08-30}{2}\right)$

$$= P(Z \leq -1.96) = 0.0250$$

322. (C)

$$P(122 < \bar{X} < 124) = P\left(\frac{122-120}{30/\sqrt{225}} < Z < \frac{124-120}{30/\sqrt{225}}\right)$$

$$= P\left(\frac{122-120}{2} < Z < \frac{124-120}{2}\right) = P(1 < Z < 2) = 0.1359$$

323. (B)

$$P(\bar{X} \le 126) = P\left(Z \le \frac{126-120}{15/\sqrt{25}}\right) = P\left(Z \le \frac{126-120}{3}\right) = P(Z \le 2) = 0.9772$$

324. (A)

$$P(1{,}200 < \bar{X} < 1{,}250) = P\left(\frac{1200-1200}{250/\sqrt{100}} < Z < \frac{1250-1200}{250/\sqrt{100}}\right)$$

$$= P\left(\frac{1200-1{,}200}{25} < Z < \frac{1250-1200}{25}\right) = P(0 < Z < 2) = 0.4772$$

325. (B)

$$P(\bar{X} \ge 92.5) = P\left(Z \ge \frac{92.5-100}{15/\sqrt{16}}\right) = P\left(Z \ge \frac{92.5-100}{3.75}\right) = P(Z \ge -2) = 0.9772$$

326. (C)

$$P(500 < \bar{X} < 520) = P\left(\frac{500-500}{100/\sqrt{200}} < Z < \frac{520-500}{100/\sqrt{200}}\right)$$

$$= P\left(\frac{500-500}{7.0710\ldots} < Z < \frac{520-500}{7.0710\ldots}\right) = P(0 < Z < 2.83) = 0.4977$$

327. (D)

$$P(\bar{X} \le 16.9) = P\left(Z \le \frac{16.9-16}{1.5/\sqrt{4}}\right) = P\left(Z \le \frac{16.9-16}{0.75}\right) = P(Z \le 1.2) = 0.8849$$

328. (A) $\bar{x} \pm z_{\alpha/2}\frac{\sigma}{\sqrt{n}} = 25 \pm 1.96\frac{8}{\sqrt{50}} = (22.8,\ 27.2)$ is the 95% confidence interval (CI). *Estimation* is the statistical process in which sample statistics are used to estimate population parameters. A *confidence interval* (CI) is a range of values believed to include an unknown population parameter. Associated with the interval is a level of confidence (such as 95%) that the interval does indeed contain the parameter of interest.

A $(1-\alpha)100\%$ confidence interval for μ when σ is known and sampling is done from a normal population or with a large sample is given by

$$\bar{x} \pm z_{\alpha/2}\frac{\sigma}{\sqrt{n}} = (\bar{x} - E, \bar{x} + E),$$

where

$(1-\alpha)100\%$ = the level of confidence
$\bar{x}$ = the mean of the sample
σ = the standard deviation of the population

n = the sample size

$z_{\alpha/2}$ = the critical value

$z_{\alpha/2}\dfrac{\sigma}{\sqrt{n}} = E$, the margin of error

When you construct a $(1-\alpha)100\%$ confidence interval for μ, you can say that you are $(1-\alpha)100\%$ confident that μ lies between $1-E$ and $1+E$.

Confidence coefficients used in real-life applications usually range from 90 to 99. When σ is known and sampling is from a normal population or $n \geq 30$, the z-distribution is the correct distribution to use for confidence intervals. Use $z_{\alpha/2} = 1.645$ for a 90% CI, 1.96 for a 95% CI, and 2.576 for a 99% CI. In general, $z_{\alpha/2}$ is the $\left(1-\dfrac{\alpha}{2}\right)$th percentile for the standard normal distribution, where $(1-\alpha)$ is the confidence level (in decimal form).

329. (B) $\bar{x} \pm z_{\alpha/2}\left(\dfrac{\sigma}{\sqrt{n}}\right) = 26.7 \pm 1.645\left(\dfrac{6}{\sqrt{9}}\right) = (23.41,\ 29.99)$ is the 90% CI.

330. (C) $\bar{x} \pm z_{\alpha/2}\left(\dfrac{\sigma}{\sqrt{n}}\right) = 301 \pm 1.645\left(\dfrac{37}{\sqrt{100}}\right) = (294.9,\ 307.1)$ is the 90% CI.

331. (D) $\bar{x} \pm z_{\alpha/2}\left(\dfrac{\sigma}{\sqrt{n}}\right) = 58 \pm 2.576\left(\dfrac{15}{\sqrt{36}}\right) = (51.6,\ 64.4)$ is the 99% CI.

332. (B) $\bar{x} \pm z_{\alpha/2}\left(\dfrac{\sigma}{\sqrt{n}}\right) = 123 \pm 1.645\left(\dfrac{25}{\sqrt{225}}\right) = (120.3,\ 125.7)$ is the 90% CI.

333. (C) The health science researcher can be 95% confident that the true average amount of fat grams in the signature sandwich of the local deli is between 16.2 and 17.8 fat grams.

To obtain this solution with the TI-84 Plus, press **STAT** and use the right-arrow key to highlight **TESTS**:

```
EDIT CALC TESTS
1:Z-Test...
2:T-Test...
3:2-SampZTest...
4:2-SampTTest...
5:1-PropZTest...
6:2-PropZTest...
7↓ZInterval...
```

Using the down-arrow key, scroll down until you see items 7 through B (these items are the confidence interval functions):

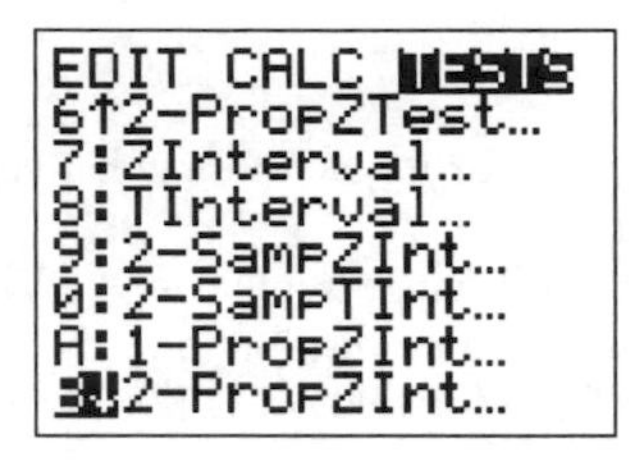

Press 7 to select the **ZInterval** function. To work with summary data, select **STATS** by pressing ENTER. Next, at the prompts, enter the given summary data: **2.5** (σ), **17** ($\bar{x}$), and **40** (n). At the **C-Level** prompt, enter **0.95** (in decimal form). The following display shows the entered data:

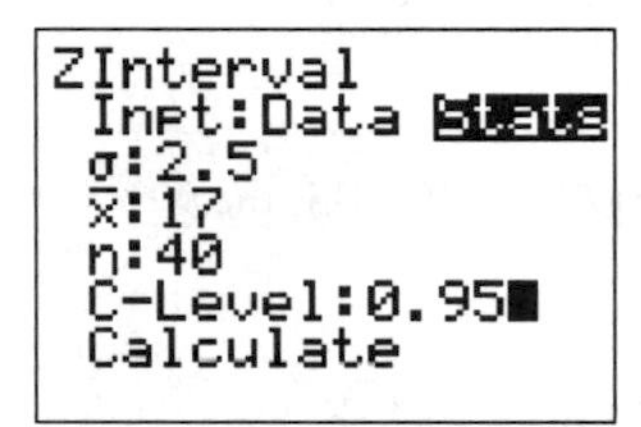

Then use the down-arrow key to scroll to **Calculate** and press ENTER. The display shows the solution:

ZInterval
(16.225,17.775)
x̄=17
n=40

334. (D) Refer to the answer to question 333 for the calculator steps. Using **ZInterval** yields (219.19, 250.51) as the 90% CI. Thus, the fund-raising officer can be 90% confident that the true average donation from contributors to the charity is between \$219.19 and \$250.51.

335. (C) Refer to the answer to question 333 for the calculator steps. Using **ZInterval** yields (36.521, 41.479) as the 95% CI. Thus, the manufacturer can be 95% confident that the true average life of its batteries is between 36.5 and 41.5 months.

336. (B) Refer to the answer to question 333 for the calculator steps. Using **ZInterval** yields (2318.5, 2481.5) as the 95% CI. Thus, with 95% confidence the true mean caloric intake of all adults in the United States is between 2,318.5 and 2,481.5 calories.

337. (C) Refer to the answer to question 333 for the calculator steps. Using `ZInterval` yields (111,482, 128,518) as the 80% CI. Thus, the real estate broker can be 80% confident that the true average house price in the metropolitan area is between \$111,482 and \$128,518.

338. (A) When σ is unknown (and the population is normally distributed), the t-distribution with $n-1$ degrees of freedom is the correct distribution to use for confidence intervals. Thus, the 90% CI is found with the following formula:

$$\bar{x} \pm t_{\alpha/2}\left(\frac{s}{\sqrt{n}}\right) = 450 \pm 1.753\left(\frac{60}{\sqrt{16}}\right) = (423.705,\ 476.295) \approx (423.7,\ 476.3).$$

Notice that s, the sample standard deviation, is used in the formula. The value for $t_{\alpha/2}$ is from Appendix C, a table that shows t critical values for degrees of freedom 1 to 32 and then shows values for selected degrees of freedom beyond 32. To use the table, first read down the column labeled "Degrees of Freedom" to locate $\text{df} = n - 1$. Then read across that row to the entry in the t_c column under "t_c Area in One Tail" corresponding to the desired confidence level. Use column $t_{.050}$ for a 90% CI, $t_{.025}$ for a 95% CI, and $t_{.005}$ for a 99% CI. In general for confidance levels, $c\ \frac{\alpha}{2} = \frac{(1 - \text{confidence level in decimal form})}{2}$.

339. (B) $\bar{x} \pm t_{\alpha/2}\left(\frac{s}{\sqrt{n}}\right) = 450 \pm 1.860\left(\frac{60}{\sqrt{9}}\right) = (412.8,\ 487.2)$ is the 90% CI.

340. (C) $\bar{x} \pm t_{\alpha/2}\left(\frac{s}{\sqrt{n}}\right) = 450 \pm 1.711\left(\frac{60}{\sqrt{25}}\right) = (429.468,\ 470.532) \approx (429.5,\ 470.5)$ is the 90% CI.

341. (A) $\bar{x} \pm t_{\alpha/2}\left(\frac{s}{\sqrt{n}}\right) = 450 \pm 3.182\left(\frac{60}{\sqrt{4}}\right) = (354.54,\ 545.46) \approx (354.5,\ 545.5)$ is the 95% CI.

342. (B) $\bar{x} \pm t_{\alpha/2}\left(\frac{s}{\sqrt{n}}\right) = 450 \pm 2.064\left(\frac{60}{\sqrt{25}}\right) = (425.232,\ 474.768) \approx (425.2,\ 474.8)$ is the 95% CI.

343. (C) $\bar{x} \pm t_{\alpha/2}\left(\frac{s}{\sqrt{n}}\right) = 450 \pm 2.947\left(\frac{60}{\sqrt{16}}\right) = (405.795,\ 494.205) \approx (405.8,\ 494.2)$ is the 99% CI.

344. (D) $\bar{x} \pm t_{\alpha/2}\left(\frac{s}{\sqrt{n}}\right) = 450 \pm 2.797\left(\frac{60}{\sqrt{25}}\right) = (416.436,\ 483.564) \approx (416.4,\ 483.6)$ is the 99% CI.

345. (B) Using `TInterval` yields (449.04, 510.96) as the 95% CI. Thus, the researcher can be 95% confident that the true average score on the standardized exam for all test takers is between 449 and 511.

To obtain this solution with the TI-84 Plus, press STAT and use the right-arrow key to highlight **TESTS**. Press **8** to select the **TInterval** function. To work with summary data, select **STATS** by pressing ENTER. Next, at the prompts, enter the given summary data: **480** ($\bar{x}$), **75** (S_x), and **25** (n). At the **C-Level** prompt, enter **0.95** (in decimal form). The following display shows the entered data:

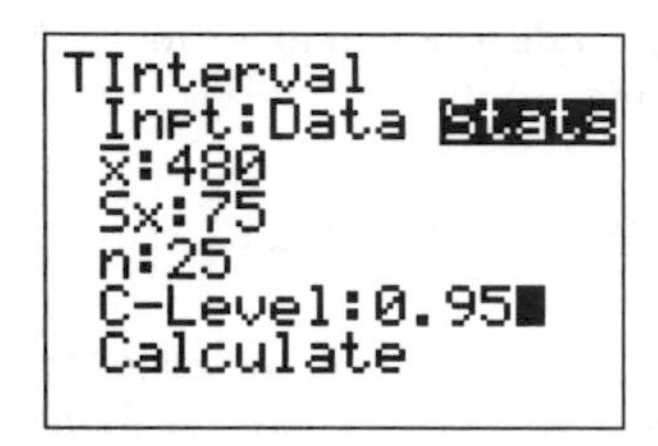

Then use the down-arrow key to scroll to **Calculate** and press ENTER. The display shows the solution:

```
TInterval
 (449.04,510.96)
 x̄=480
 Sx=75
 n=25
```

346. (B) Refer to the answer to question 345 for the calculator steps. Using **TInterval** yields (8.882, 13.118) as the 90% CI. Thus, the nutritionist can be 90% confident that the true average amount of fat grams in "heart-healthy" sandwiches is between 8.9 and 13.1.

347. (C) Refer to the answer to question 345 for the calculator steps. Using **TInterval** yields (128.54, 171.46) as the 95% CI. Thus, the statistician can be 95% confident that the true average length of time for installing a bathtub by the company is between 128.5 and 171.5 minutes.

348. (A) Refer to the answer to question 345 for the calculator steps. Using **TInterval** yields (89.882, 100.12) as the 99% CI. Thus, the hospital administrator can be 99% confident that the true mean weight of all newborn baby girls born at the administrator's hospital is between 89.9 and 100.1 ounces.

349. (D) Refer to the answer to question 345 for the calculator steps. Using **TInterval** yields (457.92, 582.08) as the 95% CI. Thus, with 95% confidence, the true mean lifetime of all incandescent light bulbs of this brand is between 457.9 and 582.1 hours.

350. (C) An attribute is a characteristic that an entity either does or does not possess. Attributes are measured as the proportion, denoted p, of the population with the characteristic in question.

The formula for a $(1-\alpha)100\%$ confidence interval for p is

$$\hat{p} \pm z_{\alpha/2}\sqrt{\frac{\hat{p}\hat{q}}{n}} = (\hat{p}-E, \hat{p}+E),$$

where

$\hat{p} = \frac{x}{n}$ = the proportion in a random sample of n entities in which x have the characteristic in question

$\hat{q} = 1-\hat{p}$

$z_{\alpha/2}$ = the critical value corresponding to a $(1-\alpha)100\%$ confidence interval

$z_{\alpha/2}\sqrt{\frac{\hat{p}\hat{q}}{n}} = E$, the margin of error

Substituting into the formula, the 90% confidence interval is

$$\hat{p} \pm z_{\alpha/2}\sqrt{\frac{\hat{p}\hat{q}}{n}} = 0.2 \pm 1.645\sqrt{\frac{(0.2)(0.8)}{200}} = (0.1535,\ 0.2465).$$

351. (A)

$$\hat{p} \pm z_{\alpha/2}\sqrt{\frac{\hat{p}\hat{q}}{n}} = 0.45 \pm 2.576\sqrt{\frac{(0.45)(0.55)}{100}} = (0.3218,\ 0.5782)$$

352. (B) First, calculate $\hat{p}$, $\hat{p} = \frac{x}{n} = \frac{45}{1500} = 0.03$

$$\hat{p} \pm z_{\alpha/2}\sqrt{\frac{\hat{p}\hat{q}}{n}} = 0.03 \pm 1.96\sqrt{\frac{(0.03)(0.97)}{1500}} = (0.0214,\ 0.0386)$$

353. (C) Apply the general rule of thumb that the sample size is large enough when both $n\hat{p}$ and $n(1-\hat{p})$ are greater than 5. Choice C is correct because $n\hat{p} = (15)(0.6) = 9$ and $n(1-\hat{p}) = (15)(0.4) = 6$, so both are greater than or equal to 5. Eliminate choice A because $n\hat{p} = (20)(0.2) = 4$, so n is too small. Eliminate choice B because $n\hat{p} = (100)(0.03) = 3$, so n is too small. Eliminate choice D because $n(1-\hat{p}) = (150)(0.01) = 1.5$, so n is too small.

354. (D)

$$E = z_{\alpha/2}\sqrt{\frac{\hat{p}\hat{q}}{n}} = 1.645\sqrt{\frac{(0.6)(0.4)}{15}} = 0.2081$$

355. (A)

$$E = z_{\alpha/2}\sqrt{\frac{\hat{p}\hat{q}}{n}} = 1.96\sqrt{\frac{(0.82)(0.18)}{1000}} = 0.0238$$

356. (A) Using `1-PropZInt` yields (0.01505, 0.04495) as the 95% CI.

To obtain this solution with the TI-84 Plus, press STAT and use the right-arrow key to highlight **TESTS**. Use the down-arrow key to scroll to **A : 1-PropZInt**. Press ENTER to select the **1-PropZInt** function. Next, at the prompts, enter the given sample data: **15** (x) and **500** (n). At the **C-Level** prompt, enter **0.95** (in decimal form). The following display shows the entered data:

```
1-PropZInt
 x:15
 n:500
 C-Level:.95
 Calculate
```

Then use the down-arrow key to scroll to **Calculate** and press ENTER. The display shows the solution:

```
1-PropZInt
 (.01505,.04495)
 p̂=.03
 n=500
```

Notice that to use **1-PropZInt**, you must enter x, the number of entities with the characteristic. If you are given $\hat{p}$ (and not x) to obtain x, use $x = \hat{p}n$.

357. (B) Refer to the answer to question 356 for the calculator steps. Using **1-PropZInt** yields (0.04717, 0.25283) as the 99% CI.

358. (C) Refer to the answer to question 356 for the calculator steps. Using **1-PropZInt** yields (0.26208, 0.41792) as the 90% CI.

359. (D) Refer to the answer to question 356 for the calculator steps. First calculate $x = \hat{p}n = (80\%)(300) = 240$. Using **1-PropZInt** yields (0.75474, 0.84526) as the 95% CI.

360. (A) Refer to the answer to question 356 for the calculator steps. Using **1-PropZInt** yields (0.01727, 0.08273) as the 90% CI.

361. (C)

$$n = \left(\frac{z_{\alpha/2} \cdot \sigma}{E}\right)^2 = \left(\frac{1.96 \cdot 10}{2}\right)^2 = 96.04$$

Thus, the minimum sample size is 97.

362. (A)

$$n=\left(\frac{z_{\alpha/2}\cdot\sigma}{E}\right)^2=\left(\frac{1.96\cdot 100}{2}\right)^2=9604$$

Thus, the minimum sample size is 9,604.

363. (B)

$$n=\left(\frac{z_{\alpha/2}\cdot\sigma}{E}\right)^2=\left(\frac{1.96\cdot 100}{0.2}\right)^2=960400$$

Thus, the minimum sample size is 960,400.

364. (C)

$$n=\left(\frac{z_{\alpha/2}\cdot\sigma}{E}\right)^2=\left(\frac{1.96\cdot 5}{1.5}\right)^2\approx 42.7$$

Thus, the minimum sample size is 43.

365. (C)

$$n=\left(\frac{z_{\alpha/2}\cdot\sigma}{E}\right)^2=\left(\frac{1.96\cdot 5}{0.15}\right)^2\approx 4268.4$$

Thus, the minimum sample size is 4,269.

366. (A)

$$n=\frac{z_{\alpha/2}^2\cdot pq}{E^2}=\frac{(1.96)^2(0.1)(0.9)}{(0.02)^2}\approx 864.4$$

Thus, the minimum sample size is 865.

367. (B)

$$n=\frac{z_{\alpha/2}^2\cdot pq}{E^2}=\frac{(1.96)^2(0.2)(0.8)}{(0.02)^2}\approx 1536.6$$

Thus, the minimum sample size is 1,537.

368. (D)

$$n=\frac{z_{\alpha/2}^2\cdot pq}{E^2}=\frac{(1.96)^2(0.5)(0.5)}{(0.02)^2}=2401$$

Thus, the minimum sample size is 2,401.

369. (A)

$$n = \frac{z_{\alpha/2}^2 \cdot pq}{E^2} = \frac{(1.96)^2(0.7)(0.3)}{(0.02)^2} \approx 2016.8$$

Thus, the minimum sample size is 2,017.

370. (B)

$$n = \frac{z_{\alpha/2}^2 \cdot pq}{E^2} = \frac{(1.96)^2(0.9)(0.1)}{(0.02)^2} \approx 864.4$$

Thus, the minimum sample size is 865.

371. (B)

$$n = \left(\frac{z_{\alpha/2} \cdot \sigma}{E}\right)^2 = \left(\frac{2.576 \cdot 7.6}{1.5}\right)^2 \approx 170.3$$

Thus, the minimum sample size is 171.

372. (D) If no prior information is available, use $p = 0.5$.

$$n = \frac{z_{\alpha/2}^2 \cdot pq}{E^2} = \frac{(1.96)^2(0.5)(0.5)}{(0.04)^2} = 600.25$$

Thus, the minimum sample size is 601.

373. (A)

$$n = \left(\frac{z_{\alpha/2} \cdot \sigma}{E}\right)^2 = \left(\frac{1.645 \cdot 8}{2}\right)^2 \approx 43.3$$

Thus, the minimum sample size is 44.

374. (B)

$$n = \left(\frac{z_{\alpha/2} \cdot \sigma}{E}\right)^2 = \left(\frac{1.96 \cdot 95}{20}\right)^2 \approx 86.7$$

Thus, the minimum sample size is 87.

375. (C)

$$n = \frac{z_{\alpha/2}^2 \cdot pq}{E^2} = \frac{(1.645)^2(0.1)(0.9)}{(0.01)^2} \approx 2435.4$$

Thus, the minimum sample size is 2,436.

Chapter 11: One-Sample Hypothesis Tests of Means and Proportions

376. (C) Because the hypotheses statements are made *before* a sample is taken, no information from the sample is used in the statements. Eliminate choice A because hypotheses concern population *parameters*, not *statistics*. Eliminate choice B because a test of hypothesis tests the null hypothesis. Eliminate choice D because the null hypothesis normally contains equality (no difference).

377. (C) When H_a contains $\neq$, the hypothesis test is two-tailed. The statements in choices A, B, and D are false.

378. (B) A test of hypothesis is significant if its *p*-value is less than α, the significance level. Because a *p*-value that is less than 1% is also less than 5%, a test of hypothesis that is significant at the 1% significance level would also be significant at the 5% level. The statements in choices A, C, and D could be true for a particular test of hypothesis but are not always true. *Caution:* The significance level should be decided before the *p*-value is calculated.

379. (B) If the *p*-value of a hypothesis test is less than α, its significance level, the null hypothesis should be rejected. The statements in choices A, C, and D are false.

380. (A) The one-sample *t*-test is the best choice for three reasons: (1) the test concerns a population mean, (2) σ is unknown, and (3) the population is normally distributed.

381. (B) The one-sample *z*-test is the best choice for three reasons: (1) the test concerns a population mean, (2) σ is known, and (3) the sample size is large ($n = 100 \geq 30$).

382. (D) None of the tests listed is appropriate, because (1) the test concerns a population mean, (2) σ is unknown, and (3) the sample size is small ($n = 4 < 30$) without the assurance that the population is normally distributed.

383. (C) The one-sample proportion *z*-test is the best choice for two reasons. First, the test concerns a population proportion. Second, both $n(p_0) = 80(0.18) = 14.4$ and $n(1 - p_0) = 80(0.82) = 65.6$ are greater than or equal to 5.

384. (B) The one-sample *z*-test is the best choice, because the test concerns a population mean, σ is known, and the sample size is large ($n = 50 \geq 30$).

385. (C) The one-sample proportion *z*-test is the best choice for two reasons. First, the test concerns a population proportion. Second, both $n(p_0) = 300(0.75) = 225$ and $n(1 - p_0) = 300(0.25) = 75$ are greater than or equal to 5.

386. (A) The one-sample *t*-test is the best choice because the test concerns a population mean, σ is unknown, and the population is normally distributed.

387. (C) The one-sample proportion z-test is the best choice because the test concerns a population proportion and because both $n(p_0) = 500(0.05) = 25$ and $n(1 - p_0) = 500(0.95) = 475$ are greater than or equal to 5.

388. (B) The one-sample z-test is the best choice because the test concerns a population mean, σ is known, and the sample size is large ($n = 500 \geq 30$).

389. (C) The one-sample proportion z-test is the best choice because the test concerns a population proportion and because both $n(p_0) = 100(0.40) = 40$ and $n(1 - p_0) = 100(0.60) = 60$ are greater than or equal to 5.

390. (D) None of the tests listed is appropriate because the test concerns a population proportion but $n(p_0) = 100(0.03) = 3$ is less than 5.

391. (A) Formulating the null and alternative hypotheses is a challenging part of a test of a hypothesis. One very important guideline is that *no information from the sample* is used in the hypotheses statements. Another helpful guideline is that the *null hypothesis always contains equality* (=, ≤, or ≥). You begin by identifying in the problem an assertion about the population parameter and then translating the assertion into symbolism. Next, you write the negation of this first assertion. Finally, you specify H_0 as the assertion containing equality and H_a as the remaining assertion. If H_a contains a "≠" symbol, the hypothesis test is two-tailed. If H_a contains a ">" symbol, the hypothesis test is right-tailed, and if H_a contains a "<" symbol, the hypothesis test is left-tailed. (Note: Some authors use only the equal symbol in the null hypothesis statement.)

In this question, the assertion about the population parameter is "the average donation from contributors to the charity is $250.00." Omitting the units, in symbols this assertion is $\mu = 250.00$. The negation of this assertion is $\mu \neq 250.00$. Now, because H_0 always contains equality, the hypotheses are as given in choice A ($H_0 : \mu = 250.00$; $H_a : \mu \neq 250.00$). Eliminate choice B because H_0 contains ≠. Eliminate choices C and D because they contain information from the sample.

392. (B) The assertion about the population parameter is "the average life of its batteries is at least 45 months." Omitting the units, in symbols this assertion is $\mu \geq 45$. The negation of this assertion is $\mu < 45$. Now, because H_0 always contains equality, the hypotheses are as given in choice B ($H_0 : \mu \geq 45$; $H_a : \mu < 45$). Eliminate choices A and C because they contain information from the sample. Eliminate choice D because H_0 contains <.

393. (C) The assertion about the population parameter is "the true mean caloric intake of all adults in the United States is no more than 2,300 calories per day." Omitting the units, in symbols this assertion is $\mu \leq 2{,}300$. The negation of this assertion is $\mu > 2{,}300$. Now, because H_0 always contains equality, the hypotheses are as given in choice C (H_0 : $\mu \leq 2{,}300$; $H_a : \mu > 2{,}300$). Eliminate choices A and B because they contain information from the sample. Eliminate choice D because H_0 contains >.

394. (D) The assertion about the population parameter is the claim that the "'heart- healthy' sandwich has a mere 10 fat grams." Omitting the units, in symbols this assertion is $\mu = 10$. The negation of this assertion is $\mu \neq 10$. Now, because H_0 always contains equality, the hypotheses are as given in choice D ($H_0 : \mu = 10$; $H_a : \mu \neq 10$). Eliminate choices A and B because they contain information from the sample. Eliminate choice C because H_a contains < instead of ≠.

395. (A) The assertion about the population parameter is "the average price of a house in a metropolitan area is $300,000." Omitting the units, in symbols this assertion is $\mu = 300{,}000$. The negation of this assertion is $\mu \neq 300{,}000$. Now, because H_0 always contains equality, the hypotheses are as given in choice A ($H_0 : \mu = 300{,}000$; $H_a : \mu \neq 300{,}000$). Eliminate choices B and D because they contain information from the sample. Eliminate choice C because H_a contains > instead of ≠.

396. (B) The assertion about the population parameter is "the mean weight of all newborn baby girls born at the hospital is less than 100 ounces." Omitting the units, in symbols this assertion is $\mu < 100$. The negation of this assertion is $\mu \geq 100$. Now, because H_0 always contains equality, the hypotheses are as given in choice B ($H_0 : \mu \geq 100$; $H_a : \mu < 100$). Eliminate choice A because the direction of the inequality symbol in H_0 should be reversed. Eliminate choices C and D because they contain information from the sample.

397. (C) The assertion about the population parameter is "the proportion of [the manufacturer's] lighting cables that are defective must be less than 0.05." In symbols this assertion is $p < 0.05$. The negation of this assertion is $p \geq 0.05$. Now, because H_0 always contains equality, the hypotheses are as given in choice C ($H_0 : p \geq 0.05$; $H_a : p < 0.05$). Eliminate choices A and B because 15 is not a proportion. Eliminate choice D because H_a should contain < instead of ≠.

398. (C) The assertion about the population parameter is "the true proportion of working women over the age of 60 in the metropolitan area is no more than 0.18." In symbols this assertion is $p \leq 0.18$. The negation of this assertion is $p > 0.18$. Now, because H_0 always contains equality, the hypotheses are as given in choice C ($H_0 : p \leq 0.18$; $H_a : p > 0.18$). Eliminate choices A and B because 80 is not a proportion. Eliminate choice D because H_0 contains >.

399. (A) The assertion about the population parameter is "the percent of radio listeners in the area who listen to the manager's radio station is at least 40%." In symbols this assertion is $p \geq 0.40$. The negation of this assertion is $p < 0.40$. Now, because H_0 always contains equality, the hypotheses are as given in choice A ($H_0 : p \geq 0.40$; $H_a : p < 0.40$). Eliminate choice B because H_a contains equality. Eliminate choices C and D because they contain information obtained from the sample.

400. (C) The assertion about the population parameter is "the proportion of assisted-living residents who prefer blue for their bedroom walls is greater than 0.75." In symbols this assertion is $p > 0.75$. The negation of this assertion is $p \leq 0.75$. Now, because H_0 always contains equality, the hypotheses are as given in choice C ($H_0 : p \leq 0.75$; $H_a : p > 0.75$). Eliminate choices A and D because H_0 contains no equality. Eliminate choice B because H_a contains < instead of >.

401. (D) The researcher made a type II error, the mistake of accepting a false H_0.

402. (A) The researcher made the correct decision to accept a true H_a.

403. (C) The researcher made a type I error, the mistake of rejecting a true H_0.

404. (B) The probability of making a type I error and the probability of making a type II error are inversely related for a fixed sample size. In some situations, the statement in choice A could be true. The statements in choices C and D are false.

405. (A) For $\alpha = 0.05$, a left-tailed test concerning μ from a normally distributed population with known σ and small $n = 20$ (< 30) has only one critical point $-z_\alpha = -z_{0.05} = -1.645$, where z_α is the $(1-\alpha)$th percentile for the standard normal distribution. H_0 will be rejected if and only if the test statistic is less than -1.645.

406. (D) For $\alpha = 0.05$, a two-tailed test concerning p has two critical points $\pm z_{\alpha/2} = \pm z_{0.025} = \pm 1.96$, where $z_{\alpha/2}$ is the $\left(1 - \frac{\alpha}{2}\right)$th percentile for the standard normal distribution. H_0 will be rejected if and only if the test statistic is less than -1.96 or greater than 1.96.

407. (C) For $\alpha = 0.05$, a right-tailed test concerning μ with unknown σ and large $n = 35$ (≥ 30) has only one critical point $t_{0.05}(34 \text{ df}) = 1.691$. H_0 will be rejected if and only if the test statistic is greater than 1.691.

408. (A) The appropriate test is the one-sample z-test because (1) the test concerns a population mean, (2) σ is known, and (3) the sample size is large (≥ 30). The formula is $T.S. = z = \frac{\bar{x} - \mu_0}{\sigma/\sqrt{n}}$, where μ_0 is the null-hypothesized value for the population parameter μ. Substituting into this formula gives

$$T.S. = z = \frac{\bar{x} - \mu_0}{\sigma/\sqrt{n}} = \frac{25 - 22}{8/\sqrt{50}} \approx 2.652.$$

409. (B) The appropriate test is the one-sample t-test because (1) the test concerns a population mean, (2) σ is unknown, and (3) the population is normally distributed. The formula is $T.S. = t = \frac{\bar{x} - \mu_0}{s/\sqrt{n}}$. Substituting into this formula gives

$$T.S. = t = \frac{\bar{x} - \mu_0}{s/\sqrt{n}} = \frac{26.7 - 28}{6/\sqrt{40}} \approx -1.370.$$

410. (C) The appropriate test is the one-sample t-test because (1) the test concerns a population mean, (2) σ is unknown, and (3) the sample size is large (≥ 30). Substituting into the formula gives

$$T.S. = t = \frac{\bar{x} - \mu_0}{s/\sqrt{n}} = \frac{301 - 300}{37/\sqrt{100}} \approx 0.270.$$

411. (D) The appropriate test is the one-sample z-test because (1) the test concerns a population mean, (2) σ is known, and (3) the sample size is large (≥ 30). Substituting into the formula gives

$$T.S. = z = \frac{\overline{x} - \mu_0}{\sigma/\sqrt{n}} = \frac{58-50}{15/\sqrt{36}} = 3.200.$$

412. (A) The appropriate test is the one-sample z-test because (1) the test concerns a population mean, (2) σ is known, and (3) the sample size is large (≥ 30). Substituting into the formula gives

$$T.S. = z = \frac{\overline{x} - \mu_0}{\sigma/\sqrt{n}} = \frac{123-120}{25/\sqrt{225}} = 1.800.$$

413. (B) $T.S. = z \approx -1.591$. To obtain this solution with a TI-84 Plus, press STAT and use the right-arrow key to highlight **TESTS**:

```
EDIT CALC TESTS
1:Z-Test...
2:T-Test...
3:2-SampZTest...
4:2-SampTTest...
5:1-PropZTest...
6:2-PropZTest...
7↓ZInterval...
```

Items 1 through 6 are the hypothesis test functions. Press 1 to select the **Z - Test** function. To work with summary data, select **STATS** by pressing ENTER. The null hypothesis is H_0 : $\mu = 250.00$ (see question 391), so enter 250.00 at the μ_0 prompt. Next, enter **95.23** (σ), **234.85** ($\overline{x}$), and **100** (n). The alternative hypothesis is $H_a : \mu \neq 250.00$, so for the μ option, select $\neq \mu_0$. The following screen shows the entered data:

```
Z-Test
 Inpt:Data Stats
 μ0:250
 σ:95.23
 x̄:234.85
 n:100
 μ:≠μ0 <μ0 >μ0
 Calculate Draw
```

Next, use the down-arrow key to scroll to **Calculate** and press ENTER. The display shows the solution:

```
Z-Test
 μ≠250
 z=-1.590885225
 p=.1116354091
 x̄=234.85
 n=100
```

Thus, $T.S. = z = -1.590885225 \approx -1.591$.

414. (C) $T.S. = t \approx -6.708$.

To obtain this solution with a TI-84 Plus, press STAT and use the right-arrow key to highlight **TESTS**. Press 2 to select the **T-Test** function. To work with summary data, select **STATS** by pressing ENTER. The null hypothesis is $H_0 : \mu \geq 45$ (see question 392), so enter 45 at the μ_0 prompt. Next, enter **39** ($\bar{x}$), **8** (Sx), and **80** (n). The alternative hypothesis is $H_a : \mu < 45$, so for the μ option, select $< \mu_0$. The following screen shows the entered data:

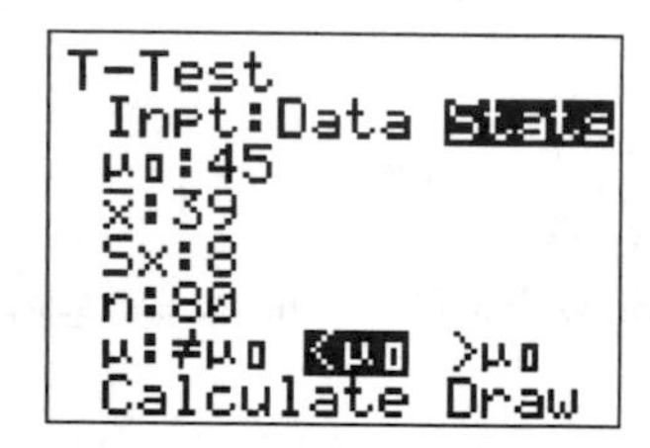

Next, use the down-arrow key to scroll to **Calculate** and press ENTER. The display shows the solution:

```
T-Test
 μ<45
 t=-6.708203932
 p=1.3427E-9
 x̄=39
 Sx=8
 n=80
```

Thus, $T.S. = t = -6.708203932 \approx -6.708$.

415. (D) Refer to the answer to question 413 for the calculator steps. The hypotheses are $H_0 : \mu \leq 2{,}300$; $H_a : \mu > 2{,}300$ (see question 393). Using **Z-Test** yields $T.S. = z \approx 2.404$.

416. (A) Refer to the answer to question 413 for the calculator steps. The hypotheses are $H_0 : \mu = 300{,}000$; $H_a : \mu \neq 300{,}000$ (see question 395). Using **Z-Test** yields $T.S. = z = 3.009$.

417. (B) Refer to the answer to question 414 for the calculator steps. The hypotheses are $H_0 : \mu \geq 100$; $H_a : \mu < 100$ (see question 396). Using **T-Test** yields $T.S. = t = -3.194$.

418. (B) A test of hypothesis is significant if its p-value is less than α, the significance level. Because $p\text{-value} = 0.0499 \leq 0.05$, H_0 will be rejected.

419. (C) Refer to the answer to question 413 for the calculator steps. The hypotheses are $H_0 : \mu = 22$; $H_a : \mu \neq 22$. Using **Z-Test** yields $p\text{-value} \approx 0.0080$.

420. (D) Refer to the answer to question 414 for the calculator steps. The hypotheses are $H_0 : \mu \geq 28$; $H_a : \mu < 28$. Using `T-Test` yields p-value ≈ 0.0892.

421. (A) See question 419. The hypotheses are $H_0 : \mu = 22$; $H_a : \mu \neq 22$. The p-value $= 0.0080 \leq 0.05$, so reject H_0. Therefore, at the 5% significance level, the data provide sufficient evidence to indicate that the mean of the population is not 22.

422. (C) See question 420. The hypotheses are $H_0 : \mu \geq 28$; $H_a : \mu < 28$. The p-value $= 0.0892 > 0.01$, so fail to reject H_0. Therefore, at the 1% significance level, the hypothesis that the mean of the population is less than 28 is not supported by the data.

423. (B) The hypotheses are $H_0 : \mu = 60$; $H_a : \mu \neq 60$. For $\alpha = 0.05$, a two-tailed test concerning μ from a normally distributed population with unknown σ and small $n = 16$ (< 30) has two critical points $\pm t_{\alpha/2}(15 \text{ df}) = \pm t_{0.025}(15 \text{ df}) = \pm 2.131$. Therefore, the decision rule is to reject H_0 if either $T.S. < -2.131$ or $T.S. > 2.131$.

424. (A) The hypotheses are $H_0 : \mu \geq 28$; $H_a : \mu < 28$. For $\alpha = 0.05$, a left-tailed test concerning μ from a normally distributed population with known σ and small $n = 25$ (< 30) has only one critical point $-z_\alpha = -z_{0.05} = -1.645$. Therefore, the decision rule is to reject H_0 if $T.S. < -1.645$.

425. (A) The hypotheses are $H_0 : \mu \leq 300$; $H_a : \mu > 300$. For $\alpha = 0.10$, a one-tailed test concerning μ from a normally distributed population with unknown σ and small $n = 15$ (< 30) has only one critical point $t_\alpha(14 \text{ df}) = t_{0.100}(14 \text{ df}) = 1.345$. Therefore, the decision rule is to reject H_0 if $T.S. > 1.345$.

426. (C) The hypotheses are $H_0 : \mu \leq 50$; $H_a : \mu > 50$. For $\alpha = 0.01$, a right-tailed test concerning μ from a population with known σ and large $n = 36$ (≥ 30) has only one critical point $z_\alpha = z_{0.01} = 2.326$. Therefore, the decision rule is to reject H_0 if $T.S. > 2.326$.

427. (A) The hypotheses are $H_0 : \mu \leq 120$; $H_a : \mu > 120$. For $\alpha = 0.05$, a one-tailed test concerning μ from a normally distributed population with unknown σ and small $n = 20$ (< 30) has only one critical point $t_\alpha(19 \text{ df}) = t_{0.050}(19 \text{ df}) = 1.729$. Therefore, the decision rule is to reject H_0 if $T.S. > 1.729$.

428. (B) The hypotheses are $H_0 : \mu = 500$; $H_a : \mu \neq 500$. For $\alpha = 0.05$, a two-tailed test concerning μ from a normally distributed population with known σ and small $n = 25$ (< 30) has two critical point $\pm z_{\alpha/2} = \pm z_{0.025} = \pm 1.96$. Therefore, the decision rule is to reject H_0 if either $T.S. < -1.96$ or $T.S. > 1.96$.

429. (D) The decision rule cannot be determined from the information given. Neither the z- nor the t-distribution applies to this situation because the test concerns a population mean, σ is not given, and the sample size is small ($n = 4 < 30$) without the assurance that the population is normally distributed.

430. (B) The hypotheses are $H_0 : \mu \leq 120$; $H_a : \mu > 120$. (Note that 2 hours = 120 minutes.) For $\alpha = 0.10$, a right-tailed test concerning μ from a population with known σ and large $n = 40$ (≥ 30) has only one critical point $z_\alpha = z_{0.10} = 1.282$. Therefore, the decision rule is to reject H_0 if $T.S. > 1.282$.

431. (B) The hypotheses are $H_0 : \mu \geq 100$; $H_a : \mu < 100$. For $\alpha = 0.01$, a one-tailed test concerning μ from a normally distributed population with unknown σ and small $n = 20$ (< 30) has one critical point $-t_{\alpha}(19 \text{ df}) = -t_{0.010}(19 \text{ df}) = -2.539$. Therefore, the decision rule is to reject H_0 if *T.S.* < -2.539.

432. (D) The decision rule cannot be determined from the information given. Neither the *z*- nor the *t*-distribution applies to this situation because the test concerns a population mean, σ is not given, and the sample size is small ($n = 5 < 30$) without the assurance that the population is normally distributed.

433. (A) Choice A is correct because $np_0 = (500)(0.05) = 25$ and $n(1 - p_0) = (500)(0.95) = 475$ are both greater than or equal to 5, so n is large enough. Eliminate choice B because $np_0 = (300)(0.01) = 3$, so n is too small. Eliminate choice C because $n(1 - p_0) = (10)(0.25) = 2.5$, so n is too small. Eliminate choice D because $n(1 - p_0) = (2{,}000)(0.002) = 4$, so n is too small.

434. (D) In choice D, because $n(1 - p_0) = (120)(0.04) = 4.8 < 5$, n is too small. Both n and $n(1 - p_0)$ are greater than or equal to 5 in choices A, B, and C.

435. (C) The hypotheses are $H_0 : p \geq 0.05$; $H_a : p < 0.05$. For $\alpha = 0.01$, a left-tailed test concerning p with sample size $n = 500$ such that $n(p_0) = 500(0.05) = 25$ and $n(1 - p_0) = 500(0.95) = 475$ are greater than or equal to 5 has only one critical point $-z_{\alpha} = -z_{0.01} = -2.326$. Therefore, the decision rule is to reject H_0 if *T.S.* < -2.326.

436. (A) The hypotheses are $H_0 : p \leq 0.18$; $H_a : p > 0.18$. For $\alpha = 0.05$, a right-tailed test concerning p with sample size $n = 80$ such that $n(p_0) = 80(0.18) = 14.4$ and $n(1 - p_0) = 80(0.82) = 65.4$ are greater than or equal to 5 has only one critical point $z_{\alpha} = z_{0.05} = 1.645$. Therefore, the decision rule is to reject H_0 if *T.S.* > 1.645.

437. (D) It cannot be determined from the information given because $n(1 - p_0) = (200)(0.02) = 4 < 5$, so n is too small.

438. (A) The hypotheses are $H_0 : p \geq 0.40$; $H_a : p < 0.40$. For $\alpha = 0.05$, a left-tailed test concerning p with sample size $n = 100$ such that $n(p_0) = 100(0.40) = 40$ and $n(1 - p_0) = 100(0.60) = 60$ are greater than or equal to 5 has only one critical point $-z_{\alpha} = -z_{0.05} = -1.645$. Therefore, the decision rule is to reject H_0 if *T.S.* < -1.645.

439. (B) The hypotheses are $H_0 : p \leq 0.75$; $H_a : p > 0.75$. For $\alpha = 0.05$, a right-tailed test concerning p with sample size $n = 300$ such that $n(p_0) = 300(0.75) = 225$ and $n(1 - p_0) = 300(0.25) = 75$ are greater than or equal to 5 has only one critical point $z_{\alpha} = z_{0.05} = 1.645$. Therefore, the decision rule is to reject H_0 if *T.S.* > 1.645.

440. (C) The hypotheses are $H_0 : p \geq 0.10$; $H_a : p < 0.10$. For $\alpha = 0.10$, a left-tailed test concerning p with sample size $n = 120$ such that $n(p_0) = 120(0.10) = 12$ and $n(1 - p_0) = 120(0.90) = 108$ are greater than or equal to 5 has only one critical point $-z_{\alpha} = -z_{0.10} = -1.282$. Therefore, the decision rule is to reject H_0 if *T.S.* < -1.282.

441. **(D)** The formula for the test statistic is $T.S. = z = \dfrac{\hat{p} - p_0}{\sqrt{\dfrac{p_0(1-p_0)}{n}}}$, where p_0 is the null-hypothesized value for the population parameter p. Substituting into this formula gives

$$\hat{p} = \frac{15}{500} = 0.03;\ T.S. = z = \frac{\hat{p} - p_0}{\sqrt{\dfrac{p_0(1-p_0)}{n}}} = \frac{0.03 - 0.05}{\sqrt{\dfrac{0.05(0.95)}{500}}} \approx -2.052.$$

442. **(A)**

$$\hat{p} = \frac{30}{80} = 0.375;\ T.S. = z = \frac{\hat{p} - p_0}{\sqrt{\dfrac{p_0(1-p_0)}{n}}} = \frac{0.375 - 0.18}{\sqrt{\dfrac{0.18(0.82)}{80}}} \approx 4.540$$

443. **(B)** The hypotheses are $H_0 : p \geq 0.40$; $H_a : p < 0.40$ (see question 399). Using **`1-PropZTest`** yields $T.S. = z \approx -1.225$.

To obtain this solution with a TI-84 Plus, press **STAT** and use the right-arrow key to highlight **TESTS**. Press **5** to select the **`1-PropZTest`** function. To work with summary data, select **STATS** by pressing **ENTER**. The null hypothesis is $H_0 : p \geq 0.40$ (see question 399), so enter 0.40 at the p_0 prompt. Next, enter **34** (x), and **100** (n). The alternative hypothesis is $H_a : p < 0.40$, so for the `prop` option, select **< p_0**. The following screen shows the entered data:

```
1-PropZTest
 p0:.4
 x:34
 n:100
 prop≠p0 <p0 >p0
 Calculate Draw
```

Next, use the down-arrow key to scroll to **`Calculate`** and press **ENTER**. The display shows the solution:

```
1-PropZTest
 prop<.4
 z=-1.224744871
 p=.1103357448
 p̂=.34
 n=100
```

Thus, $T.S. = z = -1.224744871 \approx -1.225$.

444. **(A)** Refer to the answer to question 443 for the calculator steps. The hypotheses are $H_0 : p \leq 0.75$; $H_a : p > 0.75$ (see question 400). Using **`1-PropZTest`** yields $T.S. = z = 2.000$.

445. (D) Refer to the answer to question 443 for the calculator steps. The hypotheses are $H_0: p \geq 0.10$; $H_a: p < 0.10$. Using `1-PropZTest` yields $T.S. = z \approx -1.826$.

446. (B) Refer to the answer to question 443 for the calculator steps. The hypotheses are $H_0: p \geq 0.40$; $H_a: p < 0.40$ (see question 443). Using `1-PropZTest` yields p-value ≈ 0.1103. Because the p-value is greater than $\alpha = 0.05$, it is not significant.

447. (D) Refer to the answer to question 443 for the calculator steps. The hypotheses are $H_0: p \leq 0.75$; $H_a: p > 0.75$ (see question 444). Using `1-PropZTest` yields p-value ≈ 0.0228. Because $p \leq \alpha = 0.05$, it is significant.

448. (C) Refer to the answer to question 443 for the calculator steps. The hypotheses are $H_0: p \geq 0.10$; $H_a: p < 0.10$. Using `1-PropZTest` yields p-value ≈ 0.0339. Because $p \leq \alpha = 0.05$, it is significant.

Chapter 12: Two-Sample Hypothesis Tests of Means and Proportions

449. (A) The two-sample independent t-test is the best choice because (a) the test concerns the comparison of two population means using independent random samples from the two populations where (b) the variances of the two sampled populations are equal and (c) each sample is drawn from a normally distributed population.

450. (B) The paired t-test is the best choice because the test concerns paired pre- and post-data where the population of differences between the pairs of values is normally distributed.

451. (D) None of the tests listed is appropriate because the test concerns paired data from two populations without the assurance that the population of differences between the pairs of values is normally distributed.

452. (C) The two-sample proportion z-test is the best choice because the test concerns the comparison of two population proportions using independent random samples from the two populations where the sample sizes are sufficiently large, that is, $np \geq 5$ and $n(1-p) \geq 5$ are satisfied for each sample size.

453. (D) None of the tests listed is appropriate because the test concerns the comparison of two population means using independent random samples from the two populations where the variances of the two sampled populations are equal without the assurance that each sample is drawn from a normally distributed population.

454. (B) The paired t-test is the best choice because the test concerns paired pre- and post-data where the population of differences between each pair of values is normally distributed.

455. (C) The two-sample proportion z-test is the best choice because the test concerns the comparison of two population proportions using independent random samples from the two populations where the sample sizes are sufficiently large, that is, $np \geq 5$ and $n(1-p) \geq 5$ are satisfied for each sample size.

456. (D) None of the tests listed is appropriate because the test concerns the comparison of two population proportions using independent random samples from the two populations without the assurance that the sample sizes are sufficiently large.

457. (B) The paired t-test is the best choice because the test concerns paired data from two populations where the population of differences between the pairs of values is normally distributed.

458. (C) In symbols the assertion that "the true mean Internet sale amount exceeds the true mean mail-order sale amount" is $\mu_1 > \mu_2$. The negation of this assertion is $\mu_1 \leq \mu_2$. Now, because H_0 always contains equality, the hypotheses are as given in choice C ($H_0: \mu_1 \leq \mu_2$; $H_a: \mu_1 > \mu_2$).

459. (D) The assertion that "the intervention produces higher post-assessment scores" means that $\mu_1 - \mu_2 = \mu_D < 0$. The negation of this assertion is $\mu_D \geq 0$. Now, because H_0 always contains equality, the hypotheses are as given in choice D ($H_0: \mu_D \geq 0$; $H_a: \mu_D < 0$).

460. (B) In symbols the assertion that there is "no difference" in the two population means is $\mu_1 = \mu_2$. The negation of this assertion is $\mu_1 \neq \mu_2$. Now, because H_0 always contains equality, the hypotheses are as given in choice B ($H_0: \mu_1 = \mu_2$; $H_a: \mu_1 \neq \mu_2$).

461. (C) In symbols the assertion that "adults 25 to 39 years old are less likely to prefer blue than are adults 40 to 59 years old" is $p_1 < p_2$. The negation of this assertion is $p_1 \geq p_2$. Now, because H_0 always contains equality, the hypotheses are as given in choice C ($H_0: p_1 \geq p_2$; $H_a: p_1 < p_2$).

462. (A) The assertion that "the low-fat one-month diet is effective" means the starting weight is greater than the ending weight; that is, $\mu_1 - \mu_2 = \mu_D > 0$. The negation of this assertion is $\mu_D \leq 0$. Now, because H_0 always contains equality, the hypotheses are as given in choice A ($H_0: \mu_D \leq 0$; $H_a: \mu_D > 0$).

463. (B) In symbols, the assertion "the prevalence of smoking among male adults 18 years or older in the United States exceeds the prevalence of smoking among female adults 18 years or older" is $p_1 > p_2$. The negation of this assertion is $p_1 \leq p_2$. Now, because H_0 always contains equality, the hypotheses are as given in choice B ($H_0: p_1 \leq p_2$; $H_a: p_1 > p_2$).

464. (A) The assertion of "no difference" means $\mu_1 = \mu_2$; that is, $\mu_1 - \mu_2 = \mu_D = 0$. The negation of this assertion is $\mu_D \neq 0$. Now, because H_0 always contains equality, the hypotheses are as given in choice A ($H_0 : \mu_D = 0; H_a : \mu_D \neq 0$).

465. (D) The most commonly used test statistic (*T.S.*) for the two-sample independent *t*-test under the assumption of equal population variances (see question 449) is

$$T.S. = t = \frac{(\bar{x}_1 - \bar{x}_2) - (\mu_1 - \mu_2)}{\sqrt{s_p^2\left(\frac{1}{n_1} + \frac{1}{n_2}\right)}}$$

with $df = n_1 + n_2 - 2$

where,

$$s_p^2 = \frac{(n_1 - 1)s_1^2 + (n_2 - 1)s_2^2}{n_1 + n_2 - 2} = \text{pooled variance}$$

Substituting into the formula yields $df = n_1 + n_2 - 2 = 15 + 10 - 2 = 23$

$$s_p^2 = \frac{(n_1 - 1)s_1^2 + (n_2 - 1)s_2^2}{n_1 + n_2 - 2} = \frac{(14)(18.75)^2 + (9)(14.25)^2}{23} = 293.4538$$

$$T.S. = t = \frac{(\bar{x}_1 - \bar{x}_2) - (\mu_1 - \mu_2)}{\sqrt{s_p^2\left(\frac{1}{n_1} + \frac{1}{n_2}\right)}} = \frac{(86.40 - 75.20) - (0)}{\sqrt{293.4538\left(\frac{1}{15} + \frac{1}{10}\right)}} \approx 1.601$$

Note: The value of $\mu_1 - \mu_2$ in the test statistic typically is set to zero (under the assumption of equality in the null hypothesis).

466. (A) The test statistic for the paired *t*-test (see question 450) is

$$T.S. = t = \frac{\bar{x}_D - \mu_{D_0}}{\frac{s_D}{\sqrt{n_D}}}$$

with $df = n_D - 1$

where,

n_D = number of matched pairs

$\bar{x}_D$ = sample mean difference between n_D matched pairs

s_D = sample standard deviation of n_D differences of matched pairs

To apply this formula, first compute the sample differences $x_1 - x_2 = x_D$:

Student ID Number	**1**	**2**	**3**	**4**	**5**	**6**	**7**	**8**
Pre-assessment score	44	55	25	54	63	38	31	34
Post-assessment score	55	68	40	55	75	52	49	48
Differences	−11	−13	−15	−5	−12	−14	−18	−14

Then calculate the mean $\overline{x}_D$ and standard deviation s_D of the differences on a TI-84 Plus calculator. Enter the differences as list **`L1`**, and then use **`1-Var Stats L1`** to do the computations: $\overline{x}_D = -12.25$; $s_D \approx 5.007$.

$$\text{df} = n_D - 1 = 8 - 1 = 7$$

Substituting into the formula yields

$$T.S. = t = \frac{\overline{x}_D - 0}{s_D / \sqrt{n_D}} = \frac{-12.25 - 0}{5.007 / \sqrt{8}} \approx -6.920.$$

Note: The value of μ_{D_0} in the test statistic typically is set to zero (under the assumption of equality in the null hypothesis).

467. (C) The test statistic for the two-sample proportion z-test is

$$T.S. = z = \frac{\left(\frac{x_1}{n_1} - \frac{x_2}{n_2}\right) - (p_1 - p_2)}{\sqrt{\frac{x_1 + x_2}{n_1 + n_2}\left(1 - \frac{x_1 + x_2}{n_1 + n_2}\right)\left(\frac{1}{n_1} + \frac{1}{n_2}\right)}} = \frac{(\hat{p}_1 - \hat{p}_2) - (p_1 - p_2)}{\sqrt{\overline{p}(1 - \overline{p})\left(\frac{1}{n_1} + \frac{1}{n_2}\right)}}$$

where,

x_1 = the number of entities in sample 1 with the characteristic of interest

x_2 = the number of entities in sample 2 with the characteristic of interest

$\hat{p}_1 = \frac{x_1}{n_1}$ = the proportion of sample 1 with the characteristic of interest

$\hat{p}_2 = \frac{x_2}{n_2}$ = the proportion of sample 2 with the characteristic of interest

$\overline{p} = \frac{x_1 + x_2}{n_1 + n_2}$ = the proportion in the two samples combined with the characteristic of interest.

Substituting into this formula yields

$$\hat{p}_1 = \frac{292}{430} = 0.6791$$

$$\hat{p}_2 = \frac{180}{245} = 0.7347$$

$$\bar{p} = \frac{292+180}{430+245} = \frac{472}{675} = 0.6993$$

$$T.S. = z = \frac{(\hat{p}_1 - \hat{p}_2) - (p_1 - p_2)}{\sqrt{\bar{p}(1-\bar{p})\left(\frac{1}{n_1} + \frac{1}{n_2}\right)}} = \frac{(0.6791 - 0.7347) - (0)}{\sqrt{(0.6993)(0.3007)\left(\frac{1}{430} + \frac{1}{245}\right)}} \approx -1.515.$$

Note: The value of $p_1 - p_2$ in the test statistic typically is set to zero (under the assumption of equality in the null hypothesis).

468. (B) 0.6562 N.S. The hypotheses are $H_0 : \mu_1 = \mu_2$; $H_a : \mu_1 \neq \mu_2$ (see question 460). To obtain the solution with a TI-84 Plus, press STAT and use the right-arrow key to highlight **TESTS**. Press 4 to select the **2-SampTTest** function. To work with summary data, select **STATS** by pressing ENTER. Enter the summary statistics from the two samples:

```
2-SampTTest
 Inpt:Data Stats
 x̄1:29.5
 Sx1:7.4
 n1:5
 x̄2:31.6
 Sx2:8.4
↓n2:8
```

For the alternative hypothesis, select $\mu 1 \neq \mu 2$. Select **Yes** for **Pooled**. Next, use the down-arrow key to scroll to **Calculate**, and press ENTER. The output appears on two screens:

```
2-SampTTest
 μ1≠μ2
 t=-.4575529655
 p=.656178695
 df=11
 x̄1=29.5
↓x̄2=31.6
```

```
2-SampTTest
 μ1≠μ2
↑Sx1=7.4
 Sx2=8.4
 Sxp=8.05074813
 n1=5
 n2=8
```

p-value $\approx 0.6562 > \alpha = 0.05$, so it is not significant.

469. (D) 0.0219 *. The hypotheses are $H_0 : \mu_D \leq 0$; $H_a : \mu_D > 0$ (see question 462).

$$\text{df} = n_D - 1 = 6 - 1 = 5$$

To obtain the solution with a TI-84 Plus, enter the differences in **L1**. Then, using **1-Var Stats**, calculate the mean and standard deviation of the differences: $\bar{x}_D = 6.333\ldots$; $s_D = 5.785\ldots.$

Press **STAT** and use the right-arrow key to highlight **TESTS**. Press **2** to select the **T-Test** function. Notice that the correct values from **1-Var Stats L1** are already entered into x̄, Sx, and n. Because the null hypothesis has $\mu_D \le 0$, enter 0 at the μ_0 prompt. For the alternative hypothesis, select $>\mu_0$:

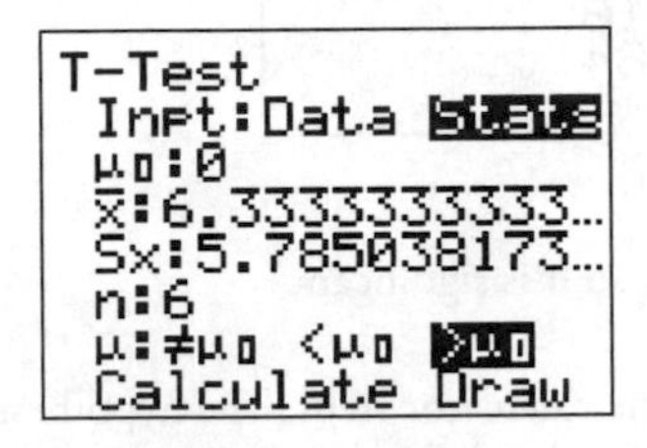

Next, use the down-arrow key to scroll to **Calculate**, and press **ENTER**. The display shows the results:

```
T-Test
 μ>0
 t=2.681647825
 p=.0218668653
 x̄=6.333333333
 Sx=5.785038173
 n=6
```

p-value $= 0.0219 \le \alpha = 0.05$, so it is significant.

470. (C) 0.0035*. The hypotheses are $H_0 : p_1 \le p_2$; $H_a : p_1 > p_2$ (see question 463). To obtain the solution with a TI-84 Plus, press **STAT** and use the right-arrow key to highlight **TESTS**. Press **6** to select the **2-SampTTest** function. Enter the statistics from the two samples. For the alternative hypothesis, select **> p2**:

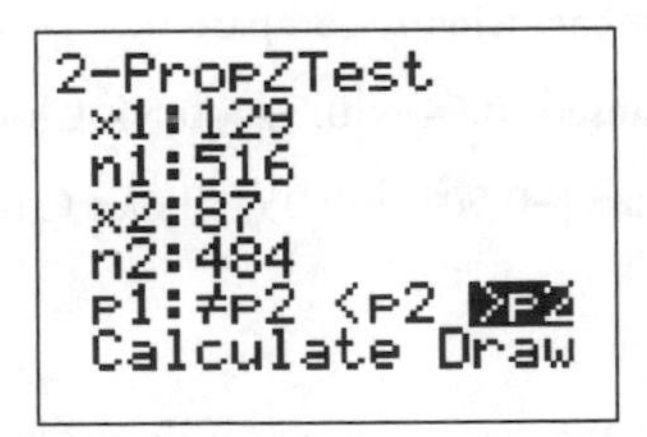

Next, use the down-arrow key to scroll to **Calculate**, and press ENTER. The output appears on two screens:

```
2-PropZTest
 p1>p2
 z=2.697715704
 p=.0034909008
 p̂1=.25
 p̂2=.1797520661
↓p̂=.216
```

```
2-PropZTest
 p1>p2
↑p̂1=.25
 p̂2=.1797520661
 p̂=.216
 n1=516
 n2=484
```

p-value $= 0.0035 \le \alpha = 0.05$, so it is significant.

471. (B) 0.8186 N.S. Refer to question 469 for the calculator steps. The hypotheses are $H_0 : \mu_D = 0$; $H_a : \mu_D \neq 0$ (see question 464). Enter the differences in **L1**. Then, using **1-Var Stats**, calculate the mean and standard deviation of the differences: $\bar{x}_D = -0.34$; $s_D = 3.104\ldots$.

Using **T-Test** yields p-value $0.8186 > \alpha = 0.05$, so it is not significant.

Chapter 13: Correlation Analysis and Simple Linear Regression

472. (D) Only the statement in choice D is always true, because Pearson product-moment correlation coefficients numerically quantify only linear relationships. The statement in choice A is true only for positive correlation coefficients. The statement in choice B is false because correlation does not imply causation. The statement in choice C is false because if two variables are independent, their correlation is zero.

473. (B) The scatterplots in choices A and B both show a linear relationship, but the scatterplot in choice B is stronger because the data values are not as spread out as those in choice A. Eliminate choices C and D because these scatterplots do not show linear relationships.

474. (D) To compare the strengths of correlation coefficients, compare their absolute values. Thus, the statement in choice D is true because $|-0.54| = |0.54| = 0.54$. Choice A is false because $|0.40| < |0.75|$. Choice B is false because $|-0.50| < |-0.95|$. Choice C is false because $|0.82| > |0.68|$.

475. (A) The correlation coefficient between W and X is the strongest relationship because it has the greatest absolute value.

476. (A) To find the correlation coefficient, –0.904, use the following formula:

$$r = \frac{n\sum xy - \left(\sum x\right)\left(\sum y\right)}{\sqrt{n\sum x^2 - \left(\sum x\right)^2}\sqrt{n\sum y^2 - \left(\sum y\right)^2}}$$

In this question, x is age, and y is pulse rate. To find n, count the number of pairs in the data set: $n = 5$. To find Σxy, multiply each x-value by its corresponding y-value, and then add the products: $\sum xy = 20 \cdot 210 + 39 \cdot 180 + 18 \cdot 200 + 44 \cdot 165 + 50 \cdot 120 = 28{,}080$. To find Σx, add the x-values: $\sum x = 20 + 39 + 18 + 44 + 50 = 171$. To find Σy, add the y-values: $\sum y = 210 + 180 + 200 + 165 + 120 = 875$. To find Σx^2, square each x-value, and then add the squares: $\sum x^2 = 20^2 + 39^2 + 18^2 + 44^2 + 50^2 = 6{,}681$. To find Σy^2, square each y-value, and then add the squares: $\sum y^2 = 210^2 + 180^2 + 200^2 + 165^2 + 120^2 = 158{,}125$.

Now substitute into the formula, and compute r:

$$r = \frac{n\sum xy - \left(\sum x\right)\left(\sum y\right)}{\sqrt{n\sum x^2 - \left(\sum x\right)^2}\sqrt{n\sum y^2 - \left(\sum y\right)^2}} = \frac{5(28{,}080) - (171)(875)}{\sqrt{5(6{,}681) - (171)^2}\sqrt{5(158{,}125) - (875)^2}}$$

$$= -0.90415 \approx -0.904.$$

477. (D) $r = 0.969$. To compute with the TI-84 Plus, enter the x-values into **L1** and their corresponding y values into **L2**, ordered so the nth elements match up as given in the data table. Press **STAT**, and use the right-arrow key to highlight **TESTS**. Then use the down-arrow key to scroll to **LinRegTTest**, and press **ENTER**:

```
LinRegTTest
 Xlist:L1
 Ylist:L2
 Freq:1
 β & ρ:≠0 <0 >0
 RegEQ:
 Calculate
```

LinRegTTest determines the coefficients of the equation, a + bx, for the best-fit line through the paired data points. At the same time, it performs a t-test on the slope and correlation and reports its p-value. The default for the alternative hypothesis is β & ρ: $\neq 0$, which essentially tests whether a significant relationship exists between the two variables. The output also includes df = degrees of freedom = $n - 2$, s = standard error about the regression line $= \sqrt{\frac{\text{Sum of }(y - \hat{y})^2}{n-2}}$, r^2 = coefficient of determination, and r = correlation coefficient.

Press ENTER ENTER to accept the default lists **L1** and **L2**. The **Freq** feature allows you to specify frequencies to points, as when a point occurs more than once, as opposed to your repeatedly listing the point. Press ENTER to accept the default of **1**. Press ENTER to accept the default alternative hypothesis β & ρ: $\neq 0$. Skipping **RegEQ** (which allows you to store the regression equation as a Y = function), use the down-arrow key to scroll to **Calculate**, and press ENTER. The output appears on two screens:

```
LinRegTTest
 y=a+bx
 β≠0 and ρ≠0
 t=7.845375367
 p=.0014258109
 df=4
↓a=47.21632124
```

```
LinRegTTest
 y=a+bx
 β≠0 and ρ≠0
↑b=7.476683938
 s=3.821935258
 r²=.9389778
 r=.9690086687
```

Thus, $r \approx 0.969$.

478. (B) Only the statement in choice B is false. The statements in choices A, C, and D are assumptions for the error terms in a simple linear regression model.

479. (D) In a simple linear regression model, the estimator for the error variance, σ^2, is mean square error (MSE):

$$\text{MSE} = \frac{\text{sum of } (y - \hat{y})^2}{n-2} = \frac{\text{sum of squares for error (SSE)}}{n-2}$$

480. (C) The statement in choice C is false because in a regression analysis, the correlation coefficient and the slope of the regression equation have the *same* arithmetic signs. The statements in choices A, B, and D are true.

481. (D) Because the coefficient of determination $r^2 = 0.81$, 81% of the variation in Y is explained by the regression relationship of Y with X. The statements in choices A, B, and C are false.

482. (C) Because the model is significant and 7 is in the range of the x-values, you can use the prediction equation to determine the predicted height (in centimeters) for a shoe size of 7: $\hat{Y} = 138.6412 + 4.4723X = 138.6412 + 4.4723(7) = 169.9473 \approx 169.9$.

483. (A) Because the coefficient of X is the slope of the regression line, for each one-unit change in the independent variable, the estimated change in the mean value of the dependent variable is -1.3041.

484. (B) The proportion of variance in Y that is statistically explained by the regression equation is $r^2 = (-0.70)^2 = 0.49$.

485. (D) It is not appropriate to use the given prediction equation for $x = 10$ hours because in regression analysis, a value of the dependent variable cannot be validly estimated if the value of the independent variable is outside the range of values used to produce the regression equation.

486. (A) Rounding the values from the output, the prediction equation is $\hat{Y} = 1.6965 + 0.2029X$. Because the model is significant and 6.5 is in the range of the x values, you can use the prediction equation to determine the predicted GPA for 6.5 hours: $Y = 1.6965 + 0.2029X = 1.6965 + 0.2029(6.5) = 3.01535 \approx 3.02$.

487. (C) The test statistic has a t distribution with $n - 2$ degrees of freedom and is computed using the formula

$$T.S. = t = \frac{b_1}{s_{b_1}} = \frac{b_1}{\sqrt{\frac{\text{MSE}}{S_{xx}}}} = \frac{b_1}{\sqrt{\frac{\text{SSE}/(n-2)}{S_{xx}}}}$$

where,

b_1 is the least-squares estimate of the regression slope β_1

s_{b_1} is the standard error of $b_1 = \sqrt{\frac{\text{MSE}}{S_{xx}}} = \sqrt{\frac{\text{SSE}/(n-2)}{S_{xx}}}$

MSE is the mean square error of the regression $= \frac{\text{SSE}}{(n-2)}$

SSE is the sum of all squared errors

S_{xx} is the sum of squares of the x's

$$T.S. = t = \frac{b_1}{s_{b_1}} = \frac{b_1}{\sqrt{\frac{\text{MSE}}{S_{xx}}}} = \frac{b_1}{\sqrt{\frac{\text{SSE}/(n-2)}{S_{xx}}}} = \frac{2.46}{\sqrt{\frac{(11.815/10)}{0.9815}}} = \frac{2.46}{1.097} \approx 2.242.$$

488. (A) The formula for a CI about β_1 is $b_1 \pm (t_{\alpha/2,(n-2)\text{ df}})(s_{b_1})$. You have $b_1 = 2.46$, $t_{\alpha/2,(n-2)\text{ df}} = t_{0.025,\,10\text{ df}} = 2.228$, and from question 487, $s_{b_1} = 1.097$. Thus, a 95% CI for β_1 is $b_1 \pm (t_{\alpha/2,(n-2)\text{df}})(s_{b_1}) = 2.46 \pm (2.228)(1.097) = (0.016, 4.904)$.

Chapter 14: Chi-Square Test and ANOVA

489. (B) The chi-square distribution is never negative. Eliminate choices A and D because the chi-square distribution assumes only nonnegative values. Unlike the z- and t-distributions, the χ^2 distribution is not symmetric, so eliminate choice C.

490. (C) Use of the chi-square statistics requires that each of the expected cell counts is at least 5.

491. (C) df = (number of rows − 1)(number of columns − 1) = $(r-1)(c-1) = (4)(3) = 12$

492. (C) Use the χ^2 table in Appendix D. Under the column labeled "Degrees of Freedom," go down to 20, and then read across until you find 34.170. The subscript on χ^2 at the top of that column is the area to the right of 34.170. Thus, the probability that a chi-square random variable with 20 degrees of freedom will exceed 34.170 is 0.025.

493. (D) The expected count in cell E_{ij}, where i is the row number and j is the column number of the cell, is given by the following formula:

$$E_{ij} = \frac{(\text{row } i \text{ total})(\text{column } j \text{ total})}{n}.$$

In the table given for this problem, the data for seniors appear in the 4th row, and the data for high anxiety appear in the 3rd column. Substitute and evaluate to find the expected number of seniors who have high mathematics anxiety:

$$E_{43} = \frac{(40)(80)}{200} = 16$$

494. (A) df $= k - 1 = 4 - 1 = 3$

495. (A) Calculate the expected value for each cell (shown in parentheses):

	Job Performance		
Sleeping Habits	**Unsatisfactory**	**Satisfactory**	**Total**
Poor	20 ($E_{11} = 12$)	10 ($E_{12} = 18$)	30
Good	20 ($E_{21} = 28$)	50 ($E_{22} = 42$)	70
Total	40	60	100

$$T.S. = \chi^2(1\text{ df}) = \sum \frac{(\text{observed} - \text{expected})^2}{\text{expected}} = \frac{(20-12)^2}{12} + \frac{(10-18)^2}{18} + \frac{(20-28)^2}{28}$$

$$+ \frac{(50-42)^2}{42} \approx 12.698$$

496. (A) Only the statement in choice A is false. The statements in choices B, C, and D are required assumptions for conducting a one-way ANOVA comparing k population means.

497. (C) In a two-way ANOVA, the significance of the interaction effect between factors 1 and 2 should be examined first.

498. (C) In an ANOVA, the variability of the observed values of the response variable around their respective treatment means is the sum of squares for treatment, $SS_{\text{treatment}}$.

499. **(B)** Use the *F*-distribution table in Appendix E. The *F*-distribution has a numerator degrees of freedom and a denominator degrees of freedom. In a one-way ANOVA, the numerator degrees of freedom is $k-1$, where k is the number of treatments, and the denominator degrees of freedom is $n-k$, where n is the sample size. Solve to find both degrees of freedom, and then use them to locate the *F*-value in the table:

$$F(k-1,\ n-k) = F(4-1,\ 20-4) = F(3,\ 16) = 3.2389$$

500. **(B)** Solve using $F(3, 16) = 3.2389$ from question 499:

$$F = \frac{\text{MST}}{\text{MSE}} = \frac{\text{SS}_{\text{treatment}} / (k-1)}{\text{SSE} / (n-k)} = \frac{763.75/3}{1{,}210/16} \approx 3.366 > F(3,\ 16) = 3.2389, \text{ so it is}$$

significant.